Health, Technology and Society

Series Editors: **Andrew Webster**, University of York, UK and **Sally Wyatt**, Royal Netherlands Academy of Arts and Sciences, The Netherlands

Titles include:

Ellen Balka, Eileen Green and Flis Henwood (*editors*)
GENDER, HEALTH AND INFORMATION TECHNOLOGY IN CONTEXT

Simone Bateman, Jean Gayon, Sylvie Allouche, Jérôme Goffette and Michela Marzano (*editors*)
INQUIRING INTO ANIMAL ENHANCEMENT
Model or Countermodel of Human Enhancement?

Simone Bateman, Jean Gayon, Sylvie Allouche, Jérôme Goffette and Michela Marzano (*editors*)
INQUIRING INTO HUMAN ENHANCEMENT
Interdisciplinary and International Perspectives

Courtney Davis and John Abraham (*editors*)
UNHEALTHY PHARMACEUTICAL REGULATION
Innovation, Politics and Promissory Science

Gerard de Vries and Klasien Horstman (*editors*)
GENETICS FROM LABORATORY TO SOCIETY
Societal Learning as an Alternative to Regulation

Alex Faulkner
MEDICAL TECHNOLOGY INTO HEALTHCARE AND SOCIETY
A Sociology of Devices, Innovation and Governance

Herbert Gottweis, Brian Salter and Catherine Waldby
THE GLOBAL POLITICS OF HUMAN EMBRYONIC STEM CELL SCIENCE
Regenerative Medicine in Transition

Roma Harris, Nadine Wathen and Sally Wyatt (*editors*)
CONFIGURING HEALTH CONSUMERS
Health Work and the Imperative of Personal Responsibility

Jessica Mesman
MEDICAL INNOVATION AND UNCERTAINTY IN NEONATOLOGY

Mike Michael and Marsha Rosengarten
INNOVATION AND BIOMEDICINE
Ethics, Evidence and Expectation in HIV

Nelly Oudshoorn
TELECARE TECHNOLOGIES AND THE TRANSFORMATION OF HEALTHCARE

Margaret Sleeboom-Faulkner
GLOBAL MORALITY AND LIFE SCIENCE PRACTICES IN ASIA
Assemblages of Life

Nadine Wathen, Sally Wyatt and Roma Harris (*editors*)
MEDIATING HEALTH INFORMATION
The Go-Betweens in a Changing Socio-Technical Landscape

Andrew Webster (*editor*)
NEW TECHNOLOGIES IN HEALTH CARE
Challenge, Change and Innovation

Andrew Webster (*editor*)
THE GLOBAL DYNAMICS OF REGENERATIVE MEDICINE
A Social Science Critique

Health, Technology and Society
Series Standing Order ISBN 978-1-4039-9131-7 hardback
(*outside North America only*)

You can receive future titles in this series as they are published by placing a standing order. Please contact your bookseller or, in case of difficulty, write to us at the address below with your name and address, the title of the series and the ISBN quoted above.

Customer Services Department, Macmillan Distribution Ltd, Houndmills, Basingstoke, Hampshire RG21 6XS, England palgrave▸pivot

Inquiring into Animal Enhancement: Model or Countermodel of Human Enhancement?

Edited by

Simone Bateman
Centre National de la Recherche Scientifique, France

Jean Gayon
Université Paris 1 Panthéon-Sorbonne, France

Sylvie Allouche
Université Catholique de Lyon, France

Jérôme Goffette
Université Claude Bernard Lyon 1, France

Michela Marzano
Université Paris Descartes, France

palgrave
macmillan

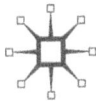

Introduction, selection and editorial matter © Simone Bateman, Jean Gayon, Sylvie Allouche, Jérôme Goffette and Michela Marzano 2015
Individual chapters © Respective authors 2015

All rights reserved. No reproduction, copy or transmission of this publication may be made without written permission.

No portion of this publication may be reproduced, copied or transmitted save with written permission or in accordance with the provisions of the Copyright, Designs and Patents Act 1988, or under the terms of any licence permitting limited copying issued by the Copyright Licensing Agency, Saffron House, 6–10 Kirby Street, London EC1N 8TS.

Any person who does any unauthorized act in relation to this publication may be liable to criminal prosecution and civil claims for damages.

The authors have asserted their rights to be identified as the authors of this work in accordance with the Copyright, Designs and Patents Act 1988.

First published 2015 by
PALGRAVE MACMILLAN

Palgrave Macmillan in the UK is an imprint of Macmillan Publishers Limited, registered in England, company number 785998, of Houndmills, Basingstoke, Hampshire RG21 6XS.

Palgrave Macmillan in the US is a division of St Martin's Press LLC, 175 Fifth Avenue, New York, NY 10010.

Palgrave Macmillan is the global academic imprint of the above companies and has companies and representatives throughout the world.

Palgrave® and Macmillan® are registered trademarks in the United States, the United Kingdom, Europe and other countries.

ISBN: 978–1–137–54248–9 EPUB
ISBN: 978–1–137–54247–2 PDF
ISBN: 978–1–137–54246–5 Hardback

A catalogue record for this book is available from the British Library.

A catalogue record for this book is available from the Library of Congress.

www.palgrave.com/pivot

DOI: 10.1057/9781137542472

Contents

Series Editors' Introduction *Andrew Webster and Sally Wyatt*	vi
Acknowledgements	viii
Notes on Contributors	x
Introduction *Simone Bateman, Jean Gayon, Sylvie Allouche, Jérôme Goffette and Michela Marzano*	1
1 Animal Enhancement: Technovisionary Paternalism and the Colonisation of Nature *Arianna Ferrari*	13
2 Improving Animals, Improving Humans: Transpositions and Comparisons *Florence Burgat*	34
3 Harming Some to Enhance Others *Gary Comstock*	49
4 Sex Hormones for Animals and Humans? Enhancement and the Public Expertise of Drugs in Post-war United States and France *Jean-Paul Gaudillière*	79
5 So Different and Yet So Similar: Comparing the Enhancement of Human and Animal Bodies in French Law *Sonia Desmoulin-Canselier*	109
Index	132

Series Editors' Introduction

Medicine, health care and the wider social meaning and management of health are undergoing major changes. In part this reflects developments in science and technology, which enable new forms of diagnosis, treatment and the delivery of health care. It also reflects changes in the locus of care and burden of responsibility for health. Today, genetics, informatics, imaging and integrative technologies, such as nanotechnology, are redefining our understanding of the body, health and disease; at the same time, health is no longer simply the domain of conventional medicine, nor the clinic. The 'birth of the clinic' heralded the process through which health and illness became increasingly subject to the surveillance of medicine. Although such surveillance is more complex, sophisticated and precise as seen in the search for 'predictive medicine', it is also more provisional, uncertain and risk laden.

At the same time, the social management of health itself is losing its anchorage in collective social relations and shared knowledge and practice, whether at the level of the local community or through state-funded socialised medicine. This individualisation of health is both culturally driven and state sponsored, as the promotion of 'self-care' demonstrates. The very technologies that redefine health are also the means through which this individualisation can occur – through 'e-health', diagnostic tests, and the commodification of restorative tissues, such as stem cells, cloned embryos and so on.

This series explores these processes within and beyond the conventional domain of 'the clinic', and asks whether they amount to a qualitative shift in the social ordering and value of medicine and health. Locating technical developments in wider socio-economic and political processes, each book discusses and critiques recent developments within health technologies in specific areas, drawing on a range of analyses provided by the social sciences.

The series has already published thirteen books that have explored many of these issues, drawing on novel, critical and deeply informed research undertaken by their authors. In doing so, the books have shown how the boundaries between the three core dimensions that underpin the whole series – health, technology and society – are changing in fundamental ways. This latest addition to the series, like its companion volume, *Inquiring into Human Enhancement: Interdisciplinary and International Perspectives*, takes this reconfiguring of boundaries further, especially as it considers the relationships between humans and other animals, and the biomedical practices that sometimes cross species.

Like its companion volume, this short collection provides an extremely rich and strongly interdisciplinary interrogation of the concept of 'enhancement'. The contributors started from the assumption that they could learn more about human enhancement by examining long-standing practices of 'improving' animals. What emerged was a set of striking differences between the practices surrounding humans and other animals. Not least, animals are being modified largely in the service of human needs and priorities, with little consideration for the animals themselves. The term 'enhancement' may not be fully appropriate as animals are manipulated or engineered in order to improve their capacities to produce food for human consumption or to serve as models for experimentation in biomedical research. As with the companion volume, this book focuses on the practices and implications of enhancement for society now and in the future.

This book is in some ways a departure for the series, through its explicit focus on animals. Nonetheless it makes a major contribution to our understanding within the social sciences and humanities of the current and likely future developments not only in animal enhancement, but also what they might mean for human enhancement and the meaning of health.

As series Editors we are delighted to mark our entrance to the Palgrave Pivot series with this provocative volume that will attract international interest from scholars working across a number of disciplines. It will also be of great interest to researchers and practitioners in biomedical fields and in animal research.

Andrew Webster and Sally Wyatt

Acknowledgements

The book you are about to read was initially part of a larger project: a series of workshops held in Paris, France, between 2009 and 2011 entitled 'Human Enhancement – A Critical Inquiry'. We brought together scholars from different countries and disciplinary environments, in an attempt to critically investigate the definition, scope and limits of this term, whose use has become so widespread that its meaning is often taken for granted. In the course of our inquiry, we devoted substantial attention to the enhancement of animals. However, because animal enhancement raises a number of complex issues, specific to these practices, a separate book on this subject seemed preferable to reducing animal enhancement to a mere prototype of the problems that human enhancement may encounter in the future. We are grateful to Palgrave Macmillan for having proposed and accepted to devote a separate book to the issues raised by animal enhancement, as a companion volume to the other book derived from our project: *Inquiring into Human Enhancement: Interdisciplinary and International Perspectives* (Basingstoke: Palgrave Macmillan, 2015).

The editors of these two volumes, who were also the organisers of these workshops, also wish to thank the Universities Paris 1 Panthéon-Sorbonne, Paris Descartes, and Paris Diderot for having financed this series of workshops. Without their generous support, we would have been unable to bring together scholars from such a wide variety of countries and disciplinary backgrounds in a concerted and constantly evolving research query on the enhancement of both humans and animals.

Acknowledgements ix

We also wish to thank the scholars who generously accepted our invitation to intervene, as speakers or as commentators, in our series of workshops: Gustaf Arrhenius, Bernadette Bensaude-Vincent, Florence Burgat, Pierre-Henri Castel, Christopher Coenen, Gary Comstock, Maxime Coulombe, Eric de Léséleuc, Sonia Desmoulin-Canselier, Françoise Dupeyron-Lafay, Selim Eskiizmirliler, Anne Fagot-Largeault, Arianna Ferrari, Marc Fleurbaey, Jean-Paul Gaudillière, Alain Giami, Bertrand Jordan, Patrick Laure, Erik Malmqvist, Anne Marcellini, Jean-Noël Missa, Marika Moisseeff, Brian Muñoz, Pascal Nouvel, Patrick Pajon, Isabelle Queval, Bernard Reber, David Rothman, Sheila Rothman, Brian Stableford, Ruud ter Meulen, Daniel Weinstock, Dan Wikler and Myriam Winance. Not all of them appear as authors in this or in the companion volume, but together with the persons who regularly took part in this series – too numerous to name here – they all contributed substantially to workshop discussions and ultimately influenced our conception of this publication project.

Last but not least, we would like to thank the heads of the collection 'Health, Technology and Society' at Palgrave Macmillan – Andrew Webster and Sally Wyatt – for having guided us through the editorial process. An equally warm expression of our appreciation goes to Harriet Barker, who proposed the publication of our project as two separate books, and to Holly Tyler and Dominic Walker, who assisted us through the complex phases of handling the publication of companion volumes, as well as to the numerous persons at Palgrave Macmillan who helped us produce this book. Most of all, we thank Andrew Webster for his unfailing support in helping us bring this publication project to fruition.

Notes on Contributors

Editors

The editors of this Pivot volume were involved in the preparation of a companion volume, published by Palgrave Macmillan, entitled *Inquiring into Human Enhancement: Interdisciplinary and International Perspectives*. Both volumes were derived from papers presented at a series of workshops on human enhancement held in Paris between 2009 and 2011.

Sylvie Allouche, PhD (Philosophy), has conducted research and taught in various European universities (Paris, Lyon, Budapest, Toulon, Bristol). She is an assistant professor (2014–15) in the Human Development Department and the General Biology Laboratory, Université Catholique de Lyon, France. Her research develops along two complementary directions: (1) the various philosophical issues raised by the prospect of engineering living organisms, with a special interest for human enhancement and geoengineering; (2) the relations between philosophy and fiction, and more specifically science fiction and television series.

Simone Bateman is Emeritus Senior Researcher in Sociology at the Centre National pour la Recherche Scientifique (CNRS); she conducts her research activities at the Centre for Research on Medicine, Science, Health, and Society (CERMES3) in Paris, France. Most of her work concerns morally controversial medical and scientific practices, primarily in the area of reproduction and sexuality (abortion, contraception, reproductive technology, neonatal intensive care), as well as other closely related practices (stem-cell research, genetic testing, experimentation on human subjects). Some of her work also concerns bioethics as a historically specific social phenomenon. She has published widely, mostly in French and English, on these topics, and has also participated in the production of reports for national and international institutions, notably a European Commission report on reproductive technology, Glover J. et al., *Fertility and the Family* (1989). She was a member of the French National Ethics Committee from 1992 to 1996.

Jean Gayon is Professor of Philosophy at Université Paris 1 Panthéon-Sorbonne, and the director of the Institute of History and Philosophy of Science and Techniques (IHPST). His work bears on the history of modern biology (evolutionary theory, genetics, biometry), philosophy of biology (concepts of species, gene, function, chance, model organism) and the history of philosophy of science. He has also published on some social and political aspects of life and health science, more especially eugenics, notion of race, conservation biology and human enhancement (list available at: http://halshs.archives-ouvertes.fr, 'gayon'). His works include a book on the history of selection theory (*Darwinism's Struggle for Survival*, 1998), 20 collective books and 270 articles or book chapters. Gayon is a member of the German National Academy of Science 'Leopoldina', the International Academy of History of Science, the International Institute of Philosophy, the International Academy of Philosophy of Science and Academia Europaea.

Jérôme Goffette is Associate Professor of Philosophy of Medicine at the Université Claude Bernard Lyon 1 and a member of the University's research unit on *Science and Society: History, Education, and Practices*. His research focuses primarily on anthropotechnics and human enhancement, a topic on which he has written more than 20 articles, as well as a book: *Naissance de l'anthropotechnie* (Birth of anthropotechnics) (2006). His second area of research concerns the body and its imaginary. Among other publications on this topic, he co-edited a collective volume with Lauric Guillaud entitled *L'imaginaire médical dans le fantastique et la science-fiction* (Medical imagination in fantastic and science fiction) (2010); and with Jonathan Simon, 'The Internal Environment: Claude Bernard's Concept and Its Representation in *Fantastic Voyage* (R. Fleisher)', in Landers M. and Muñoz B. (eds), *Anatomy and the Organization of Knowledge, 1500–1850* (2012, pp. 187–205).

DOI: 10.1057/9781137542472.0004

Michela Marzano is Professor of Philosophy at the Université Paris Descartes where she conducts research on applied ethics. Her work bears on moral norms and values (autonomy, consent, dignity), bioethics (euthanasia, use of human embryos, allocation of scarce health resources) and sexual ethics (fidelity, pornography, rape). Her works include: *G.E. Moore's Ethics. Good as Intrinsic Value* (2004); *Dictionnaire du corps* (ed., 2007); *Dictionnaire de la violence* (ed., 2011).

Contributors

Florence Burgat is a senior researcher in Philosophy at the Institut National de la Recherche Agronomique (INRA), and a statutory member of the CNRS – École Normale Supérieure research unit 'Husserl Archives'. Her research interests focus primarily on phenomenological approaches to animal life and the animal condition in industrial societies, and she has also co-hosted a research seminar at the 'Husserl Archives' on the theme of the animal agent. She is co-editor in chief of the *Revue Semestrielle de Droit Animalier*, published by the University of Limoges. Her latest works are *Liberté et inquiétude de la vie animale* (2006); *Penser le comportement animal. Contribution à une critique du réductionnisme* (edited volume) (2010); *Une autre existence. La condition animale* (2012); *Ahimsa. Violence et non-violence envers les animaux en Inde* (2014); *La cause des animaux. Pour un destin commun* (2015).

Gary Comstock is Professor of Philosophy at North Carolina State University where he conducts research on ethical questions in the biological sciences. His most recent book is *Research Ethics: A Philosophical Guide to the Responsible Conduct of Research* (2013). Of his earlier book *Vexing Nature? On the Ethical Case against Agricultural Biotechnology* (2000), one critic wrote that the volume was a watershed in the discussion of genetically modified foods. Another declared that Comstock's nuanced treatment of the issue was 'virtually unprecedented in applied philosophy'. Comstock was ASC Fellow of the National Humanities Center (2007–09) and serves as editor in chief of two online scholarly communities: *On the Human* and *Open Seminar in Research Ethics*. Comstock edited *Life Science Ethics* (2002, 2010) and *Is There a Moral Obligation to Save the Family Farm?* (1987).

Sonia Desmoulin-Canselier has a PhD in Private and Criminal Law (2005) from the Université Paris 1 Panthéon-Sorbonne. Since 2007, she is a researcher at the Centre National de la Recherche Scientifique (CNRS) and conducts her research activities at the CNRS-Université Paris I Research Unit on Comparative Law, in the team working on 'Law, Science, and Techniques'; she is also a contractual lecturer at the Université Paris 1 Panthéon-Sorbonne. She studies the relationships between science and law, that is, not only the legal framework for scientific and technical activities, but also the mutual influences between scientific knowledge and technical progress, on the one hand, and legal rules, on the other. Her main works deal with animal law (animal legal status, animal health and animal experimentation), with legal issues linked to emergent technologies (biotechnologies, nanotechnologies and neurosciences) and with legal theory (definitions, categories, interactions between scientific language and legal language etc.).

Arianna Ferrari, philosopher, is head of the research area 'Innovation Processes and Impacts of Technology' at the Institute for Technology Assessment and System Analysis (ITSA) at the Karlsruhe Institute of Technology in Germany. Her research interests focus primarily on the interface between ethics and politics of emerging technologies (in particular in the life sciences), human/animal studies and philosophy of technology. She co-edited the first technology assessment study on animal enhancement commissioned by the Swiss Federal Ethics Committee on Non-Human Biotechnology (ECNH) (Ferrari et al., *Animal Enhancement. Neue technische Möglichkeiten und ethische Fragen*) and has published in different journals on topics concerning animal experiments, animal ethics and human and animal enhancement. She recently co-edited the first handbook on human/animal relationships in German (Ferrari A. and Petrus K. *Lexikon der Mensch-Tier-Beziehungen*, Transcript, Bielefeld, in press) and is currently working on a project on *in vitro* meat, funded by the German Ministry of Education.

Jean-Paul Gaudillière is a senior researcher at the Institut National de la Santé et de la Recherche Médicale (INSERM), and director of the Centre for Research on Medicine, Science, Health, and Society (CERMES3) in Paris, France. His research was first focused on the molecularisation of biology during the 20th century and later on the reconfiguration of medical research after the Second World War. He is the author of several edited volumes on these issues and of *Inventer la Biomedicine* (2002).

He is working on the history of biological drugs before the advent of gene-based technology, with strong interests in the dynamics of knowledge production, clinical work and market construction. He has recently coordinated *Drug Trajectories* (Studies in History and Philosophy of the Biological and Biomedical Sciences, 2005) and *How Pharmaceuticals Became Patentable* (History and Technology, 2008).

Introduction

Simone Bateman, Jean Gayon, Sylvie Allouche, Jérôme Goffette and Michela Marzano

Abstract: *Can the age-old practices of animal selection and breeding and the more recent biotechnological interventions on animals, far more intrusive and systematic than any present form of human enhancement, enlighten us as to the future of enhancement practices? This book explores issues raised by past and present practices of animal enhancement in terms of their means and goals, and identifies lessons that can be learned about enhancement, as it concerns both animals and humans. The extreme ambiguity of the claim that animals are being enhanced, the driving goals and strategies of third-party interests, but also the similarity of the substances and techniques used on humans and animals, suggest that human enhancement practices may be as tainted by equivocal goals and strategies as equivalent practices with animals.*

Bateman, Simone, Jean Gayon, Sylvie Allouche, Jérôme Goffette and Michela Marzano, eds. *Inquiring into Animal Enhancement: Model or Countermodel of Human Enhancement?* Basingstoke: Palgrave Macmillan, 2015. DOI: 10.1057/9781137542472.0005.

When we first began to organise a series of interdisciplinary workshops on the concept and practices of *human enhancement* more than five years ago, the idea of preparing a book dedicated to animal enhancement was far from our minds. Our curiosity was focused on the heated intellectual debate in many countries regarding whether or not we should use biomedical and biotechnological interventions – such as genetic modification or doping – to improve human capacities and performances. Moreover, an 'intellectual and cultural movement' called *transhumanism* had emerged at the turn of the millennium: it openly promoted 'improving the human condition through applied reason, especially by developing and making widely available technologies to eliminate aging and to greatly enhance human intellectual, physical, and psychological capacities' (Bostrom et al.,, 2003, p. 4). Animal enhancement as such was not discussed during the early phases of this debate; attention was focused almost exclusively on the enhancement of humans, and concern with the 'uplifting' of animals was late to arrive on the scene (Dvorsky, 2008 [2006]).

What motivated our initial project was therefore our wish to understand whether the term *enhancement* designated totally new practices and new goals based on the advancement of science and technology or whether these were simply old practices – such as eugenics or progressivism – in new clothing. Moreover, as opposed to the dominant concern with the moral issues raised by human enhancement practices – is enhancement good and desirable or is it ultimately bad and dangerous? – we felt the need to explore a somewhat different set of issues by taking an epistemological approach to this area of concern. For what indeed is enhancement all about? What does 'the improvement of capacities' mean? The distinctive characteristic of that project, which is being published in a companion volume – *Inquiring into Human Enhancement: Interdisciplinary and International Perspectives* (Bateman et al., 2015) – was its focus on clarifying a new concept and its application in the multiple contexts within which it is currently used.

Within this perspective, one of our hypotheses was that something could be learned about enhancing human beings by examining the age-old practices of animal selection and breeding – practices that might be considered prototypical forms of animal enhancement – and comparing them to present and prospective practices with humans. Owners of domesticated animals have always tried to improve their stock by identifying those individuals bearing desired traits and then judiciously selecting and (sometimes) crossbreeding them; today, genetics and biotechnologies allow breeders to pursue this selection of desired traits by other means. Transgenesis also makes it possible to alter animal traits so as to create completely new capabilities in animals, such as that of producing pharmaceuticals, or to generate human-animal hybrids that may serve as experimental models for human disease. Moreover, multiple substances and techniques can be used to stimulate animal performance in domains as varied as horse racing and food production. Could the enhancement of animals, far more intrusive and systematic than present forms of enhancement in humans, enlighten us as to the future of enhancement practices and the problems that attempts to enhance humans might raise?

Animal enhancement is a far more complex phenomenon than commonly suspected, as the numerous references in this volume suggest. 'Animal enhancement' (as well as 'plant enhancement'), taken as conventional terms, seem to have appeared only recently, within the dual context of the emergence of novel biotechnologies and the advent of a heated debate on 'human enhancement'. Prior to this, the usual term in English was 'animal breeding', whereas in Latin languages, the expression 'animal improvement' was preferred (e.g., 'amélioration des animaux' in French, or 'mejoramiento animal' in Spanish). In all cases, these words referred mainly or exclusively to techniques aimed at improving animal stock through artificial selection and hybridisation for the benefit of human use. A radical change in the technologies available to pursue these practices and, ultimately, the emergence of the expression 'animal enhancement' have significantly affected the perception of these practices. This has generated intense polemics as to whether or not the term 'enhancement', commonly used to refer to the improvement of humans through direct technological interventions on their bodies, could be appropriately applied to such practices with animals. The possibility that animal enhancement might not be a model, but rather a countermodel for human enhancement, justified an investigation of animal enhancement in its own right.

This book is therefore devoted to the issues raised by past and present practices of animal enhancement, both in terms of their means and their goals, as they concern animals themselves, but also in their multifaceted relationship to human enhancement. In what sense can one say that an animal is being enhanced or improved, as opposed to simply being modified, manipulated, engineered or transformed into a marketable product? What does such a practice aim to improve or enhance: the animal's life, its living conditions, its appearance, its size? To what end is the animal being enhanced and, above all, to whose end? Does the term 'enhancement' still apply when the goal does not immediately serve the enhanced individual? Five scholars – three philosophers, an historian and a legal scholar – explore the heuristic potential of this concept in areas such as animal breeding, competition, food production, animal experimentation, where many innovative biotechnological practices are already presented as forms of improvement, but with no concern as to the foundation or the justifications for this claim. Our hope is that this venture will help clarify conceptual issues, identify the lessons that can be learned about enhancement practices, as they concern both animals and humans, and possibly suggest new paths for further investigation.

Whereas our quest was initially guided by a search for similarities, it is the differences between the enhancement of humans and that of animals that finally emerged as most striking. The contributions that are now assembled in this volume all abundantly illustrate the fact that practices designated today as animal enhancement lead us far astray from any notion of enhancement as an action resulting in an individual who is in any way better off, much less 'better than well', as is often claimed with respect to the enhancement of humans. A scrupulous examination of the discourse and routine practices associated with animal enhancement shows that the notion of an improved individual, stock or species requires qualification and that issues specific to the so-called improvement of animals have been systematically ignored, even when some of these practices are identified as potentially harmful for humans.

The following are the most salient issues. First and foremost, almost all of the contributors – the three philosophers Arianna Ferrari, Florence Burgat and Gary Comstock in particular – point out that, in a world where humans define the priorities for action, they also decide what the improvement of an animal means. It is almost always the interests of the owners that drive animal enhancement, and these are rarely the animal's own interests. Moreover, as pointed out by legal scholar Sonia Desmoulin-Canselier in her chapter on French law and enhancement, all civil law systems – whose framework draws heavily from Roman law – establish a legal distinction, and thus a difference in treatment, between persons and things; in this perspective, animals fall within the category of things and are thus considered the property of their owners. Animals are therefore rarely enhanced to improve their own well-being or to respond to their specific needs, but rather to further the objectives of a human enterprise, be it that of increasing food production (Ferrari, Burgat, Gaudillière); protecting investments, markets and capital assets (Comstock, Gaudillière); furthering scientific research (Ferrari, Comstock); producing prize animals for exhibits or special performances (Ferrari, Burgat); or simply providing companionship for humans (Ferrari, Burgat). In short, animal enhancement today is rarely about improving animal lives.

The distinction between persons and things notwithstanding, the techniques used to enhance animals and humans share common features that lead far beyond the traditional analogy between animal breeding and eugenics. Almost all authors – historian Jean-Paul Gaudillière and legal scholar Desmoulin-Canselier in particular – have pointed out that the substances and techniques used to enhance humans and animals are similar, when not identical, due to the fact that humans and animals share many common traits as living entities and that many of these techniques were initially experimented on non-human animals – mostly mammals – before being developed for human use. This continuum in the means used to obtain animal and human enhancement is best illustrated by the techniques associated with genetic enhancement, in particular reproductive techniques, such as artificial insemination, in vitro fertilisation, embryo selection, cloning and transgenesis. With the exception of reproductive cloning and transgenesis that have – for the time being? – been banned from human use for moral reasons, these techniques have all been used in the contexts of both animal and human reproduction. However, as pointed out by Desmoulin-Canselier, whenever human use is concerned, the civil law distinction between persons and things has made it possible to draft a more restrictive regulatory framework that reclassifies enhancement practices in humans as 'therapy'. Consequently, the similarity of the means used to 'enhance' the lives of animals and humans can in no way allow us to deduce a similarity of the goals being pursued, predetermine whose goals these might be or identify the values driving their pursuit. An understanding of the goals and values that drive the use of enhancement techniques requires greater attention to the context in which each practice is developed and to the interests that are at stake.

This is precisely why we invited our five scholars to address animal enhancement as they encounter it in their research, and consider its merits as a possible model – or as a countermodel – for human enhancement.

In the opening chapter, philosopher Arianna Ferrari examines animal enhancement from the perspective of its transhumanist advocates. She notes that deciding what technological interventions qualify as enhancements is a question that is open to interpretation, and highlights the ambiguity of the criteria used to decide this, such as those proposed by bioethicist Sarah Chan (2009, p. 679): (1) increasing a natural function or conferring a novel one; (2) improving an aspect of the animal for human purposes; (3) enabling the fulfilment of the animal's own interests. She also critically examines the visions of animal welfare that underlie contemporary technological interventions on farm animals and pets, which include practices that claim to ameliorate an animal's condition by diminishing a capacity (such as breeding blind chickens because they suffer less in crowded facilities). She points to the lack of concern for animal suffering, experimentation and exploitation in the current literature, even among transhumanists. Animal and human enhancement do share some common ground in her view, since redesigning the animal world is only a small part of a 'technovisionary' project that aims to adapt all living beings to an 'enhanced' sociotechnical environment.

In the following chapter, philosopher Florence Burgat approaches animal enhancement from a wider perspective, given that she broadens her understanding of the term to past attempts to improve animals through selective breeding. Drawing her examples from the history of the science and practices of animal husbandry in France since the 18th century, she challenges the idea that interventions on animals, whatever the context, ever leads to any real form of 'improvement': animal health, well-being and ultimately their species traits are almost never taken into consideration. In contrast to Ferrari, she questions the validity of a comparison between human and animal enhancement; in her view, these are two conceptually distinct projects because of the disregard in animal enhancement for animal interests. Both Burgat and Ferrari nonetheless agree that the utility of animals for human purposes emerges as the primary normative framework for animal enhancement.

Philosopher Gary Comstock focuses his chapter on animal experimentation, but begins with a preamble that directly responds to the question raised by this volume. We can indeed draw lessons, he states, about the future of human enhancement by looking at the methods and practices of animal breeding. As unexpected as some novel technologies may seem, past examples show that any opposition to them will eventually fade if the undesirable effects are not burdensome and if someone is making profits from this venture. Wealth will tend to become concentrated in fewer hands and public regulation will have trouble catching up. Moreover, human enhancement technologies will necessarily be experimented on animals and, if perfected, will ultimately be applied to humans. He therefore proposes a close examination of the arguments that might justify harming animals in any experiments conducted for the benefit of others. These 'others' may be humans, but he also considers examples of research that could ultimately benefit the species to which the experimental animal belongs. He examines the robustness of the arguments used both in favour and against the use of animals in such experiments, pointing to some of the weaknesses of the most extreme positions. He nonetheless concludes that research must meet the most stringent requirements with respect to the rights of animals as 'subjects-of-a-life' (Regan, 1983), before harm to experimental animals in view of any greater moral good – including that of animals themselves – may be justified.

As an historian of science, Jean-Paul Gaudillière proposes a comparative case study. He critically examines, from the dual perspective of animal and human enhancement, the public debates over sex hormone prescription, as these were played out in the United States and in France, during the 1970s and the 1980s. At that time, the same hormone – Diethylstilbestrol (DES), a synthetic oestrogen – was being liberally prescribed by physicians to pregnant women, as an off-label drug authorised for use only in women who were at risk of miscarriage and premature deliveries; it was also being widely used in the cattle industry, as a growth enhancer to increase beef and milk production. The case study highlights two distinct ways of understanding enhancement: one aimed at improving something or someone (in this case pregnancy), and the other aimed merely at augmenting a trait or capacity (growth enhancement). Gaudillière's case study reminds us that practical experiments in enhancement have often been conducted concomitantly on animals and humans; that these experiments frequently rely on the same techniques and substances, generating new risks for both animals and humans; and that these experiments have not always been successful. His case study also richly illustrates the point made by Comstock about the difficulties encountered in controlling and regulating innovation.

In the final chapter, legal scholar Sonia Desmoulin-Canselier also adopts a comparative approach in assessing the way French law differentially handles technically similar enhancement practices. Indeed, enhancement in humans and animals often relies on the same techniques because of the genetic and physiological similarity of human and animal bodies. However, the civil law distinction between persons and things makes it possible for French law to set human enhancement practices apart, usually as medical treatment. By creating what also appears as a moral boundary between human and animal practices, it becomes easier to prohibit certain practices with humans that are fully permitted with animals (or vice versa). More recently, French legislation, for example, in the area of doping, has jointly addressed the problems related to the enhancement of performance in both animals and humans – an approach that has so far not always resulted in their equal protection. Because the interests of third parties may be at stake even in human enhancement projects, the incentives to preserve – for example – the long-term health of an athlete may not be as strong as that of a prize racehorse.

This brief overview of the five chapters confirms that there remains strong disagreement as to whether or not 'enhancement' is an appropriate term to describe practices that alter traits or create new capacities in animals when there is no intention to benefit the 'enhanced' individual. Some authors, however, Gaudillière in particular, consider the term appropriate, especially if it is used by the protagonists themselves, because whatever the aim of these practices, augmentation or improvement, they share common features from which we may draw lessons. The chapters thus reveal the underlying tensions that plague any attempt to compare enhancement practices, notably with respect to the question as to whether or not a conceptual distinction between animals and humans is necessary or even pertinent. This tension is clearly what structures the arguments and counterarguments in this book.

The five contributions show that no single approach to this question can be deemed entirely satisfactory when applied to all situations. The first three chapters, written by philosophers, defend slightly distinct points of view, but all three basically argue from a perspective that safeguards animal rights and interests, and consequently in favour only of interventions that directly serve animal and not – or not only – human ends. This robust position has its weaknesses, in that the humans who devise and propose these practices are rarely the best advocates or interpreters of animal interests; nor is it evident how humans would proceed to establish animal interests unequivocally. Moreover, concern with protecting animal rights and interests may paradoxically reinforce a 'human/animal divide', by neglecting to take into consideration the similarities of the situations in which enhancements are proposed, the traits common to animals and humans, and their equal need for respect and protection. The last two chapters, written by an historian and a legal scholar, take a closer look at these similarities and explore some of the problems raised by treating animals and humans in the same way. They indeed show that attempts to improve well-being may generate risks and produce harmful side-effects on both sides of the 'divide', but also that opting to forego all distinctions may result in the neglect of differences that are significant in protecting specific groups or individuals.

Consequently, our investigation of some of the issues specific to the practice of 'enhancing' animals does not necessarily invalidate our initial hypothesis concerning the relevance of animal enhancement as a model for evaluating the problems that human enhancement may encounter in the future. In fact, each chapter spotlights the extreme ambiguity of any enhancement project – its driving goals, its current strategies and its ultimate consequences – suggesting that human enhancement may also be tainted by equivocal goals and strategies.

So, in this perspective, what other specific issues raised by animal enhancement may ultimately be significant in understanding the future of human enhancement practices?

The first issue concerns the practices that might be most frequently used to enhance humans. As regards animal enhancement, the chapters point to the frequent use of three types of practices: techniques associated with selective breeding and genetic modification; administration of pharmacological substances to modify a capacity or performance; and interventions on physical appearance. Technologies that link animal bodies to machines, such as in the case of Brain-Machine Interface technology (see Bateman *et al.*, 2015, chapter 7 by Eskiizmirliler and Goffette), are not commonly used, although they are not absent, as testified by ongoing scientific research (see note 4 in chapter 1 by Ferrari). The three most frequent practices are present across the board, in all areas of activity concerning animals, from food production to the production of animals for research, competition or for companionship. Thus, if animal enhancement is a possible model, does this suggest that human enhancement may move forward more easily – if and when it does – not so much through futuristic ventures such as human-machine hybridisation, but through projects that proceed from more familiar forms of intervention that affect everyday aspects of human life such as reproductive capacity, intellectual and physical performance and physical appearance?

This does not mean that such enhancements will be readily accepted. If we consider the case of the genetic enhancement in humans, the debate over its acceptability has a long history and has always been highly controversial. It began as far back as ancient Greece, when Plato first evoked the idea of state-controlled selective breeding for humans in his philosophical treatise *The Republic*, has since continued through the various stages of the eugenics movement in the late 19th and early 20th century and is now being played out again in the intense moral and legal debates over the appropriate use of reproductive and genetic technology. As repulsive as genetic enhancement may seem to some, this is nonetheless the well-known terrain upon which others may find it easier to embark, as opposed to ventures involving radical bodily transformation.

Another characteristic of animal enhancement is that it focuses more frequently on groups of animals – a breeder's stock, a scientist's laboratory animals – or on the improvement of a particular species. At first glance, this apparently distinguishes animal enhancement from human enhancement as it is understood today, in that the latter – at least in discourse – gives credence and worth to the expression of each person's will to be (or not to be) enhanced. But this presentation of human enhancement leaves out part of what we take for granted in analysing animal enhancement: questions about the context in which the project is undertaken. Who controls the way enhancement techniques are used? Whose goals and values prevail in the enhancement procedure: those of the person seeking to be enhanced, those of the person or institution offering the means towards that goal or a complex combination of both? What investments have been made in developing these techniques and who will gain from their diffusion? The premises that underlie specific enhancement projects may need more critical examination than is usually accorded in the debate over human enhancement.

Moreover, human enhancement projects, particularly as presented by transhumanists, often share with animal enhancement an explicit interest in an improvement of the species. Desmoulin-Canselier states, in the last chapter of this book, that the notion of human species has made its way into French law and suggests that the moral and legal distinctions between persons and things may progressively be giving way to a new way of thinking about animals and humans as a continuum of living beings, or might we say as 'subjects-of-a-life'. A common approach to animals and humans, based on their shared status as living entities – as animals, human and non-human –, may have some positive consequences, as, for example, making it easier for humans to conceive of animals as sentient beings and thus bearers of natural rights. The claim of non-human animals to moral status as persons is an alternative approach that is being widely debated (see, for example, Beauchamp and Frey, 2011, Part III). However, if one abandons, in the case of humans, the complex and sometimes unsatisfactory concept of personhood in favour of a more all-encompassing concept such as 'subject-of-a-life', might this not ultimately weaken the human claim to hard-gained rights, such as respect for autonomy, that rest on far more than the fact that humans are sentient beings? This question is open to debate, but the idea that enhancement concerns entities whose claims to rights are based essentially on their status as sentient beings could ultimately make it easier to reduce both humans and animals to mere objects of heteronomous interests.

All chapters point, in one way or another, to the importance of experimentation in the development of enhancement practices. This suggests that a particularly pertinent angle for further comparison and anticipatory reflection on the future of enhancement may possibly be a renewed examination of the history and current practices of animal and human experimentation. Indeed, experimentation appears to be a crucial crossroad where human and animal rights and interests meet, but also conflict. Both animals and humans have played a role as experimental subjects, usually for projects that aim towards advancements in therapy, but also for much of what is proposed today as an enhancement. Animals have often, but not always, preceded humans in experimental procedures, but it is true that as persons, human rights and interests are better protected than those of animals. Much has already been written about experimentation, but a dynamic comparative approach to this issue from the perspective of the development of enhancement practices would certainly produce matter for another book.

May we warmly thank the five authors, who so generously contributed their distinctive points of view to this inquiry into the past and future of animal and human enhancement practices, for having also opened so many challenging new paths for future investigation!

References

Bateman S., Gayon J., Allouche S., Goffette J. and Marzano M. (eds) (2015) *Inquiring into Human Enhancement: Interdisciplinary and International Perspectives* (Basingstoke: Palgrave Macmillan).

Beauchamp T. and Frey R. (eds) (2011) *The Oxford Handbook of Animal Ethics* (Oxford: Oxford University Press).

Bostrom N. et al. (2003) *The Transhumanist FAQ – A General Introduction* (Version 2.1), World Transhumanist Association. Downloaded on 27 April 2012: http://www.transhumanism.org/resources/FAQv21.pdf.

Chan S. (2009) 'Should we enhance animals?' *Journal of Medical Ethics*, 35, 678–83.

Dvorsky G. (2008) 'All together now: Developmental and ethical considerations for biologically uplifting nonhuman animals'. *Journal of Evolution and Technology*, 18 1: 129–42. [Conference originally presented at Stanford University in 2006.]

Regan T. (1983) *The Case for Animal Rights* (Oakland: University of California Press)

1
Animal Enhancement: Technovisionary Paternalism and the Colonisation of Nature

Arianna Ferrari

> **Abstract:** *This chapter reconstructs the debate around animal enhancement and describes what is currently being done in experimental research. It then goes on to show how visions of animal enhancement are currently discussed by their transhumanist advocates. These discussions use the positive rhetorical force of an expression like 'enhancement', bypass the practical aspects of what supporting 'animal enhancement technologies' concretely means – thus the problem of animal experiments – and rely on general arguments that stress the need to eliminate all the negative sides of 'nature'. Suggesting a strong form of human-centred paternalism, the animal enhancement project presents 'nature' as a last frontier which can be colonised by the human technological enterprise.*

Bateman, Simone, Jean Gayon, Sylvie Allouche, Jérôme Goffette and Michela Marzano, eds. *Inquiring into Animal Enhancement: Model or Countermodel of Human Enhancement?* Basingstoke: Palgrave Macmillan, 2015.
DOI: 10.1057/9781137542472.0006.

14   *Arianna Ferrari*

1   The debate on 'animal enhancement'

Originally developed to refer to medical interventions which go beyond therapeutic purposes (Lenk, 2002; President's Council of Bioethics, 2003), the concept of '*human* enhancement' was then extended to refer to science-based technological interventions on the human body aimed at improving human capabilities (European Parliament STOA, 2009).[1] The concept of '*animal* enhancement' has since emerged as a transposition of the analogous term for humans, designating technological interventions aimed explicitly at improving animal performance.[2] Bioethicist Sarah Chan (2009, p. 679) describes the enhancement of an animal as something that '(1) [p]roduces an increase in some natural function or confers a novel function; (2) [i]mproves some aspect of the animal for human purposes; (3) [e]nables greater fulfilment of the animal's own interests'.

The most quoted examples of animal enhancement research are related to the study of the mechanisms of cognitive functions, especially of learning and memory, done on animals in order to get results for humans,[3] such as genetically engineered mice that manifest improved cognitive abilities, adult mice with human neural stem cells engrafted in the brain (the 'human neuron mouse') and animals made to interact with computer interfaces[4] (Tang et al., 1999; Mamiya et al., 2003; Lehrer, 2009; see also Ferrari et al., 2010).

Research on cognitive enhancement has also reached the screen: in the movie *Rise of the Planet of the Apes*, scientists searching for a cure to Alzheimer's disease create a new breed of apes with human-like intelligence. Cognition is, however, not the only property that can be enhanced: virtually all animal properties can be modified, especially through genetic engineering. In the experimental context, transgenic animals are created through the insertion of human genes in order to study human diseases (such as 'humanized mice'; see, among others, Macchiarini et al., 2005). Other examples are: experiments aimed at creating disease- resistant transgenic animals in agriculture, in particular mastitis-resistant transgenic cows and goats (Donovan, 2005; Wall et al., 2005; Maga et al., 2006; Gottlieb and Wheeler, 2008); the creation of so-called Enviropigs™, genetically modified pigs which produce the enzyme phytase in their salivary glands so that their manure becomes less of a pollutant for the soil[5] (Forsberg et al., 2003); and applications of cloning techniques to animals used in agriculture and sport[6] (Galli et al., 2003; 2008). Furthermore, there are attempts to modify the physical appearance of animals using cosmetic surgical procedures, especially on dogs[7] (Young, 2009; Schaffer, 2009).

One of the first documents in which we can find the explicit idea of 'animal enhancement' is the so-called NBIC report on converging technologies, where the possibility of directly controlling the genetics of animals in agriculture together with the construction of nano-enabled sensors to monitor the health and nutrition of cattle is seen as part of a larger project aimed at improving human performances (Roco and Bainbridge, 2002, p. 5).[8] In his book *Citizen Cyborg* (2004), the transhumanist advocate James Hughes defends the principle of equal consideration of interests, thus considering the obligation to uplift 'disabled' animal citizens to be as strong as our obligation to uplift disabled human citizens. Hughes thinks here of technologies that could provide dolphins and chimpanzees in captivity with the means 'to think more complicated thoughts and communicate with humans, ranging from systems to translate between human speech and animal thoughts to genetic enhancements of their brains' (pp. 225–6). Another transhumanist advocate George Dvorsky (2006), director of Non-Human Persons program at the Institute for Ethics and Emerging Technologies, argues that uplift biotechnologies represents a new primary good which confronts us with the moral obligation of uplifting non-human beings and eventually including them into what has traditionally been regarded as human society. Similar arguments are used also by bioethics and legal scholar Sarah Chan (2009) and by the transhumanist philosopher Julian Savulescu (2011).

In his 'abolitionist project', the co-founder of Humanity+ (formerly known as the World Transhumanist Association) David Pearce (2007; 2011) advocates the use of technologies to eradicate suffering from the living world. This includes the reduction or elimination of suffering at the individual level through the use of microelectrodes (to directly stimulate the reward centres), neuropharmacology and gene therapy, as well as at a more general level through the biotechnological transformation of animal predators into herbivores, which he calls 'reproductive revolution'.[9] Pearce, who as a vegan opposes the human exploitation of animals, proposes to change the framework of enhancement from a human-centred perspective to a non-anthropocentric one, calling for the development and use of technologies that serve 'good' purposes, such as the eradication of suffering.

In the debate on improving animal capacities, it soon became clear that one could also think of technoscientific interventions aimed at ameliorating the conditions of animals by diminishing their abilities.

These particular interventions have since been discussed under the label of 'animal *dis*enhancement'. One of the oldest examples is the breeding of blind chickens, which apparently suffer less than normal chickens in crowded facilities. These interventions have been considered by some authors as a solution to animal welfare problems associated with crowding in the poultry industry (Thompson, 2008). Further examples can be found in genetic engineering, such as interventions aimed at knocking out the capacity to suffer and feel pain in animals in order to create insentient laboratory animals (Rollin, 1995) or insentient animals for the food industry (Henschke, 2012). Although some issues concerning animal disenhancement are peculiar to animal enhancement, such as the 'moral conundrum' of our intuitions regarding animal welfare and the need to diminish suffering (Thompson, 2008; Palmer, 2011) and to provide solutions for animal agriculture (Henschke, 2012; Schultz-Bergin, 2014), both debates show that enhancing and diminishing properties are two faces of the same coin (Ferrari, 2012b). It is indeed open to interpretation what technological interventions can be considered enhancements. This is because the very concept of 'enhancement' is not neutral, but normative, because it suggests a comparison and an act of evaluation. In order to classify an intervention as an enhancement, we need to establish criteria and goals as well as aspects with respect to which we judge these interventions. Furthermore, an improvement with respect to one criterion can also produce a deterioration with respect to a different criterion. This is clearly visible in the definition of animal enhancement suggested by Chan (2009) mentioned earlier, in which three different criteria (novelty, human interests, animal's own interests) are proposed as equally possible for animal enhancement.

In current research practices, *there* is a profound link between *animal* enhancement and *human* enhancement: animals are *de facto* the objects of exploration and change in the technoscientific age because most of the research is performed on them, including the research that will possibly lead to the development of human enhancement technologies.

This material link between human and animal enhancement reflects a profound asymmetry regarding the technological applications in these two fields: while in *human* enhancement, only human interests are considered (although interpreted differently), in the case of animal enhancement, both animal and human interests are taken as parameters by its advocates (Chan, 2009; Savulescu, 2011).

2 The silence about animal experiments

Regarding the reality of scientific practices, none of those who defend the animal enhancement project take into account the current state-of-the-art of the promoted technologies, clearly connected with the suffering and killing of billions of animals in experiments. Genetic engineering of animals is connected with the suffering and use of many (other) animals in order to breed transgenic ones, and it still relies on imprecise techniques often associated with detrimental unpredictable effects on the phenotype (Parnace et al., 2007; Ferrari, 2008; Kues and Niemann, 2014). Moreover, it is also one of the major causes for the explosion in the number of animals used in experiments (Ferrari, 2006). The case of cloning is similar: although the efficiency of this technique varies according to the animal species, cloning an animal remains a difficult task in general and is still at an experimental stage; it is mostly unpredictable in its effects on the phenotype and is profoundly linked to the human use of animals. If we take the example of cloning sport horses, the reason behind the support for cloning as an important reproductive technology lies in the current practice of sterilising these animals to render them more manageable for sport competitions. It is interesting to note that some of the major players in the American horse industry (such as the American Quarter Horse Association and the Jockey Club[10]) are opposed to the use of this technology, because they regard it as unnatural and worry about possibly compromising the animal's health (Church, 2006).

Chan (2009, p. 681), who explicitly points out the reality of animal experimentation in this case,[11] does not problematize it at all, even if she does argue that 'if we have obligations to act in animals' interests, to benefit them and not to harm them, then we have an obligation to use enhancement technologies on animals when it is in those animals' interests, and to refrain from doing so when it is against their interests. The greater the interest, the stronger the obligation in each case'. Pearce (2007; 2011), who promotes neuropharmacology, gene therapy and genetic engineering in order to abolish suffering in the living world, does not even mention the current suffering of animals for the development of these technologies. Furthermore Savulescu (2011) explicitly advocates the creation and use of genetically engineered animal models in research because of their promising potential for gaining important knowledge for the development of enhancement technologies.

Since advocates of animal enhancement generally ignore the use and abuse of billions of animals in scientific experiments, their call for fulfilling our duties toward animals appears at best ridiculous and at worst hypocritical. In their argumentation, animal enhancement advocates push speculative visions of technoscientific developments, but do not really engage with the ethical issues of current research. They refer to the general possibility of rendering animals 'better than well' but hardly consider the potential of technological interventions to protect animal's own interests. In veterinary medicine, interventions that go beyond therapeutic purposes (as the term is frequently used in the human context) do not exist outside the reference to human use of animals.[12] Dog-breeding cultures, for example, which can be considered as an expression of the desire to make dogs conform to some arbitrary physical 'ideal', constitute one of the most widespread forms of legitimization of technological, that is, surgical interventions on animals for non-therapeutic reasons connected to their use by humans (Hubrecht, 1995). Practices such as tail docking,[13] ear cropping or even de-vocalisation (in order to prevent excessive barking in dogs) are clear examples of such non-therapeutic interventions performed by veterinarians in the interests of the animal's owner. Whereas, for example, ear cropping, a common practice in the breeding of dog races such as Dobermans and Schnauzers, is now illegal in European countries because it is considered a form of maltreatment of animals, it is still performed in the USA and in some Canadian provinces.[14] It is interesting to note that the official reason is no longer aesthetic but hygienic and preventive (such as the prevention of ear infection or of tail injuries during a hunt). Veterinary interventions that go beyond therapeutic purposes, such as face-lifting on bulldogs, even when justified by the argument that accentuated wrinkles can cause bacterial infections (since the skin is not well aerated), exemplify human interventions that mitigate and correct previous human interventions on animals (such as breeding). Bulldogs 'are one of the most popular breeds according to AKC® Registration Statistics, due to their lovable and gentle dispositions and adorable wrinkles'[15]; however, at the root of the 'problem of wrinkles' is the breeding of dog races with particular characteristics for human use (fight) or for aesthetic pleasure. Even if one remains in the therapeutic context, the modern tools of veterinary medicine, such as genetic testing, are sometimes used to eliminate traits which are not desired by humans, and yet which are not connected with any particular disease in animals, such as long fur or particular fur colours (Lyons, 2010).

If we take a close look at the examples quoted from concrete research projects, we can easily notice that it is the framework of human utility which deeply informs current enhancement practices. In agriculture, the selection is oriented toward the improvement of productivity; in the breeding of dogs and cats, the selection or genetic alteration of capabilities is oriented toward diminishing their aggressiveness (and thus selecting races that are particularly well-adapted to families with children) or toward improving traits useful for other human practices, such as hunting and other sports, and so on. In current research, consequences of these interventions on animal welfare as well as on other specific traits are *only* of secondary relevance and, when they are looked for, it is always in connection with animal utility for the human being. Even if a disease like mastitis is connected with suffering, death and even mass killing,[16] and is therefore a problem for cows, it should be considered within a wider context in order to be evaluated. Genetic engineering of farm animals entails many technical difficulties; it implies the use of a huge number of other animals (which are usually killed) and is connected with unpredictable phenotypic effects that are very often detrimental to the welfare of the modified animals. These animals are often kept in isolation or under special conditions that are not welfare-friendly (Ferrari, 2008). The mass-killing of cows infected with mastitis has less to do with the prevention of suffering for the single cow, than it does with the high density of animals in breeding facilities that would favour the rapid spread of the disease. Furthermore, the mammary glands of milk cows have become more stressed, weaker and thus more susceptible to bacterial infection due to the specific breeding conditions and intensive use of milk cows. It would be wrong to say that the cause of mastitis lies in the way milk cows are used, because the real, *biological* cause is the bacterial infection. But it would also be wrong not to consider the environmental factors correlated with this disease and its changing role over time. The logic underlying the perpetuation of the framework of human use of animals is also visible in the production of the so-called Enviropigs™, where the issue is no longer that of the (mass) production of pigs for human consumption, but rather that of their transformation to render them less 'harmful' to the environment in order to continue their consumption.

Of course, we can imagine the possibility of technologies that are developed outside the framework of human utility, such as the application of sophisticated diagnostic or therapeutic tools to avoid suffering, pain and premature death of animals, as well as devices to detect adverse phenotypic traits (adverse for the well-being of the animals and not for the human being!). There are some important examples that come from the development of genetic testing for some animals both in the field of prevention and cure (Lyons, 2010), as well as the application of knowledge from genomics research to the genetic elimination of 'deleterious' genes or gene sequences in particular animal populations (mostly cats and dogs; Meyers-Wallen, 2003). However, also in these cases, ethical problems remain if the research is not conducted in a way that respects animals' fundamental rights. Indeed, genetic selection remains controversial because of the long-term unpredictability of phenotypic effects and of the loss of biodiversity (Meyers-Wallen, 2003).[17] Furthermore as already shown, in some cases, genetic testing is *de facto* performed to eliminate traits that are not problematic for the animal but for its owner (such as fur colours; see Lyons, 2010).

3 Technovisionary paternalism

The omissions about animal experiments or the explicit support of them, even materially,[18] by the majority of the advocates of animal enhancement are not very surprising on closer analysis. The transhumanist movement supports the belief in the extraordinary power of technoscientific developments for solving the great challenges of our time, since in their opinion technologies can provide the possibility of redesigning nature according to our needs and preferences. For the advocates of enhancement, the natural world, which includes the human being, is something fundamentally plastic and changeable that can be rebuilt in its very fundamental building blocks by new technologies (Ferrari, 2010; 2013). The human being, as *Homo faber*, the creator of technologies, thus holds a special responsibility in the world. Humans are the ones who can intervene on nature and start an infinite and creative process of technological modifications (Roco and Bainbridge, 2002); therefore, the enhancement of nature appears as the essence of science and technology and as a 'true' project of emancipation. Transhumanist advocates, who argue for a form of liberal eugenics (Savulescu, 2011), claim to develop the same ethical argument in the case of animals, stating that we have been ameliorating animals through breeding programmes for centuries (Chan, 2009). Indeed, there is a very close relationship between the emergence of eugenic programmes and animal breeding. With the development of genetics at the beginning of the 20th century, the modification of animal properties so as to improve their productivity has become an essential task of animal breeding, mostly in the field of agriculture. In one of the first handbooks of genetics in agriculture written by J.L. Lush, *Animal Breeding Plans* (1937), genetics is described as the science orientated to the improvement of the production of farm animals. Reconstructing the history of the American Breeders' Association (ABA), founded by agricultural scientists in 1903 in support of scientific agriculture, Kimmelmann (1983) has pointed out that it was precisely the practical interests of agricultural scientists that brought to their attention the developments in genetics (such as Mendelian, biometric and cytological studies). These interests made ABA one of the first national membership-based organisations promoting genetic and eugenic research in the United States. The selection of animal properties in agriculture was conducted with the same methods as the selection of desirable traits in human beings: therefore, the agricultural context of this association played a crucial role in the development and promotion of eugenic research in the United States at that time. Selective breeding of animals has since then implicitly worked as a form of enhancement, inasmuch as it has been guided by the idea that the modification of animal properties should be oriented towards making animals better fit different human needs.

What animal enhancement advocates add with respect to past interventions is the fact that current and future technologies seem to allow for modification of animals that makes them 'better than well'. The label 'enhancement', indeed, suggests something of an added value. However, what this 'better than well' means for the different beings involved (humans and animals) remains ambiguous. While Pearce (2007) clearly proposes the abolition of suffering as the guiding ethical principle for enhancement interventions, Chan (2009) states, without further explanation, that any improvement of the capacity of an *animal* to better fit *human* interests (see Chan, 2009) has to be considered a form of *animal enhancement*. If we applied this to the human case, I think we would hesitate to call an improvement of a *human* capacity to better fit *non-human* interests a form of *human* enhancement. Actually, this thought experiment has never been attempted.

Persson and Savulescu (2012), advocating the need for moral enhancement in the human domain, describe current human beings as 'unfit for the future' because, until now, they have proved themselves incapable of providing efficacious solutions to the problem of self-destruction due to climate change, and to the threats posed by weapons of mass destruction, especially those in the hands of terrorist groups. The way they describe human history is one of colonising Earth:

> If we look back upon the 80,000 years or so that have passed since *Homo sapiens* began to colonize the Earth starting from Eastern Africa, we discern a process of relentless expansion and exploitation, with very few episodes of restraint. (p. 100)

What is true for human nature is also true for the rest of the non-human material world. For transhumanists, the logic of current experimental research does not appear problematic but rather necessary and even positive (Savulescu, 2011). Transhumanists think that we should support technoscientific development because it is capable of offering unlimited doping for both human and non-human beings, and thus the possibility of complete control of nature. It is easy to state that transhumanism goes hand in hand with technoscientific projects that are oriented toward the systematic reprogramming of living beings when their performances are not good enough: if there are ethical, social or environmental problems associated with some forms of animal use, we should – provided the efficacy of technologies – solve the problem by transforming animals rather than by changing our use of them (Ferrari, 2012b). In other words, if we need castrated horses for sport competitions and we then have a problem reproducing them, then let us develop a technological device (cloning) to ensure their reproduction; if we have bacterial diseases in cattle that can spread very rapidly in intensive husbandry facilities, then let us engineer disease-resistant cattle! This way of thinking does not surprise at all if we consider that Liao et al. (2012) have discussed the possibility of genetically engineering humans in order to render them fit for climate change.

In defending technovisions without critically engaging with current practices in experimental research, animal enhancement advocates are convincing only if one accepts their strong human-centred paternalism: we humans know best what is good and useful for animals, even whether or not they need technologies. This is in contrast to both their appeal to our duties toward animals as well as their general liberal framework, in which individuals should be free to make their own choices. Raven (2011), referring to research aimed at rendering chimpanzees more similar to humans, called this a form 'of colonialist arrogance, and a form of species fascism'.

Although, of course, not every form of paternalism toward animals should be considered as intrinsically wrong, nor every form of intervention on animals – since we do develop cures for their diseases and different kinds of help – the paternalism of the animal enhancement project is disturbing because it justifies oppressive technologies developed at the cost of enormous suffering of billions of animals.

4 Animal enhancement as colonization of nature

The idea of enhancing nature is based on a permanent construction of the 'wrong nature' of living beings. Suddenly, it is no longer enough to be 'normal': we all have to become more powerful, faster, smarter and more beautiful. Whether one is human, animal or other, everyone and everything on Earth now appear, in different degrees and ways, 'unfit for the future' (Persson and Savulescu, 2012). The construction of the unfit nature of beings becomes more evident in the debate on animal disenhancement, where the dilemmatic nature of interventions aimed at diminishing the fundamental properties of animals works only through the presupposition that either it is impossible to change the status quo of animal exploitation (Henschke, 2012) or that animal use is not a problem (Ferrari, 2012b). The focus on animal capabilities as well as an approach which suggests that the welfare of animals can be ameliorated alternatively by augmenting or even by diminishing animal properties, fit well the logic of 'animal welfarism' that supports the protection of animals (through the maintenance of certain standards of care), but it does not question their exploitation (Gruen, 2011). For these reasons, I have called these visions forms of a 'technovisionary animal welfarism' (Ferrari, 2012a). Although Pearce calls for an abolitionist project, and thus formally rejects reformism, in my opinion, his project can be regarded as being covered by this description because of his ableist framework. Pearce closely follows philosopher Peter Singer's line of argument, which lies between an animal welfare and an animal liberationist perspective. Singer (2011) focuses on the minimisation of suffering but, due to his utilitarian framework, he accepts (harmful) animal experiments under some conditions.

In transhumanism, nature can be considered as the last frontier to be colonised, and the colonisation has been promised as a good one because it would be aimed at improvement, as a material utopia dominated by positive 'moral' values. Discourse on the 'unfitness' of nature has been criticised as embedded in a form of discrimination of particular abilities, thus as a form of ableism (Wolbring, 2009; 2010). 'Ableism', a term that evolved from the civil rights movement in the USA and in the UK during the 1960s to question the discrimination of people whose body structures and functioning abilities were labelled as 'impaired' or as sub-species typical, has since developed into a philosophical perspective that also informs the perception of certain 'enhancement technologies' (or enhancement technological visions) (Wolbring, 2008). Although the reference to abilities of individuals is indeed something that characterises our thinking as living beings, ableism has been used historically, and is still being used by various social groups, to justify their elevated level of rights and status in relation to other groups of beings, thus with a clear discriminatory purpose.

The focus on animals' productive characteristics or on their cognitive abilities (so that they can even start to better communicate with us; see Dvorsky, 2006) works only on the presupposition that animals lack something, and that they therefore need human technological interventions to improve their state. In this light, animals are evaluated, in a way comparable to humans, in connection with their performances: this reflects the fact that we live in a 'performance enhancing society', where living beings are judged on the basis of their performances (Coenen, 2010). The focus on abilities and on the transformative power of emerging technologies goes together with the capitalisation of living beings: suddenly, not only are the properties of living beings defined in economic terms (life capital, maximisation of performances and so on) but the technological transformation of the characteristics of the living being also becomes a big business through patenting and other forms of economic incentives (Cooper, 2008). The adoption of economic axioms in all areas of life, together with the belief that there is a technological fix for every problem, constitutes the essential normative assumptions of a new 'market ethics' (Harvey, 2007), common to different varieties of neoliberalism (Birch, 2008).[19]

With the idea of animal enhancement, not only is ableism projected on animals but also the idea of disability as such. This conception of nature appears at the end profoundly discriminatory and oppressive. As a matter of fact, who are we to decide on a project that would redesign nature? We do not even really know what kind of conception of 'disabilities' animals develop and how they react to that. In a group of chimpanzees in Florida, for example, it was observed that Knuckles, a chimp with cerebral palsy that rendered him incapable of behaving in a way compatible with the social hierarchy typical of chimpanzee groups, was not marginalised by the group but rather treated with special regard. Only very rarely was Knuckles subjected to intimidating displays of aggression from older male members of the group.[20] In the perspective of animal enhancement, animals are not respected in their otherness and appreciated as such; they are perceived as deficient and the human being, as the creator of technologies, knows what is best for them.

5 Conclusions

In the last half of the century, we have experienced rapid technological improvement that has offered us new possibilities for intervening on animals in ways that change their species-specific properties for implementing their utility to human beings. And yet our technologized century has also been the century in which the ethical, political and social debate concerning animals has gained importance and involved many scholars. The controversy on technoscientific development and technological visions, such as animal enhancement and disenhancement, reflects the ambivalence and tension that characterise the modern human-animal relationship: on the one hand, there are many lay people as well as scholars who are denouncing the poor conditions of animal husbandry and the disastrous consequences of animal products on the environment, and who are consequently opting for a vegan diet[21]; on the other hand, scientific research is investing money and resources in the engineering of farm animals, such as the Enviropigs™, so that they pollute less, making it possible to maintain or even to increase the actual level of animal consumption.

For sure, animals in our society, in which they are injured and killed for the protection of the health of human beings and the environment (animal experiments), for the production of food (especially in those countries where this is no longer needed), and in which they are kept under stressful conditions to satisfy different human psychological needs (such as entertainment and sports), need no other new forms of domination masked as technoscientific beneficial interventions. The promoters of the animal enhancement project do not question animal exploitation in the first place (Dvorsky, 2006; Chan, 2009; Savulescu, 2011).[22] Even Pearce, who is a vegan and proclaimed supporter of animal rights, and who criticises the suffering inflicted upon animals (except in science), ends up justifying oppressive practices in animal experiments so that we might possibly reach the paradisiac state in which there is no suffering. In a framework which permanently constructs the disability of nature and then proposes technosciences as a remedy to that, the emancipatory logic of liberating beings from something negative (something called 'bad nature', or suffering) is reversed: it cannot be perceived as a message of tolerance, respect for diversity and truly empathic living together on Earth, but only as a project of totalitarian domination.

Notes

1. Although at the beginning of the discussion, enhancement, in the human context, was identified with something intrinsically positive by the majority of its supporters (Harris, 2007), recently even strong advocates of enhancement such as Bostrom and Savulescu (2009, pp. 3–4) have chosen to accept a 'normalisation' of this concept, that is, to accept the idea that these types of interventions are definable (thus distinguishable from previous forms of human improvement like education) and are open to an evaluation (thus not necessarily positive in principle) that takes into consideration their concrete circumstances and consequences.
2. For practical reasons I will refer to non-human animals as animals in this chapter, even if the first expression is the most precise.
3. This does not surprise at all, since cognition and neurological research have gained considerable importance in the contemporary debate on *human* enhancement.
4. In research on brain-machine interfaces and prosthetics, monkeys are used to test the possibility of controlling artificial limbs (prostheses) or moving cursors on a monitor through electrodes implanted in the brain (see, e.g., Lebedev and Nicolelis, 2006). In these experiments, animals have to undergo painful surgery, as they are in many cases paralysed in order to simulate a human disease. Sometimes they are kept under painful conditions in order to force them to participate in the experiment's tasks. Furthermore, mostly in military research, there have been several attempts to remotely control animals, which could then be used to explore buildings and to carry out espionage missions. The first result of remote control was obtained in a rat in 2002 (Talwar et al., 2002), and, since then, several experiments have been performed on birds (especially pigeons) and on insects, even if many

technical difficulties must be overcome before they can be used in practice (such as, e.g., the short lifespan of insects, problems of interactions and conflicts between induced electrical stimuli and other stimuli such as lights, as well as disturbances in the electrical stimulation created by environmental factors such as wind) (Ferrari et al., 2010). Due to the secret nature of military research which is rarely published, trends in this field remain anecdotal. Even if these experiments are clearly examples of impairment (rather than improvement) of animals, robotics is often indicated as a potential enhancement technology (Dvorsky, 2006).

5   In areas of intensive pig breeding, the high concentration of phosphorus in pigs' manure is a major source of pollution, because this substance may leak into ponds, streams and rivers, causing the growth of algae and consequently the death of aquatic animals, as well as risks for the human beings who drink this water. The idea underlying the creation of Enviropigs™ is to help reduce the environmental impact of pig breeding and production, since these pigs produce less phosphorous (Golovan et al., 2001). The limited production in controlled research environments of the Enviropigs™, following a technique developed at the Canadian University of Guelph in collaboration with the industry association Ontario Pork, was approved in Canada in 2010. The research stopped in June 2012 because Ontario Pork decided to end its financial support and all Enviropigs™ were killed.

6   Cloning is per se only a technique for the production of animals that does not imply any improvement of a specific capability. However, the application of cloning in the field of agriculture and sport competitions is motivated by the concern to preserve the capabilities of those individuals which have been judged as particularly valuable sources of biological material for reproduction (see Ferrari et al., 2010).

7   Especially in the US, many clinics specialised in cosmetic surgery for pets are flourishing: they offer many different treatments, including face-lifting, dental implants (and brackets), testicular implants for dogs and ear implants. Motivations underlying this kind of treatment are varied: owners want to improve the physical appearance of their pets in order to render them more competitive in beauty contests or simply for their aesthetic enjoyment.

8   This report also refers to experiments on animals that have concretely led to an improvement of animal capacities (Roco and Bainbridge, 2002, p. 112).

9   See http://www.abolitionist.com/reprogramming/.

10 See, for example: http://www.jockeyclub.com/registry.asp?section=3#one.

11 'Technologies for human enhancement could also be applied to non-human animals—indeed, as mentioned, it is probable that many human enhancement technologies, if they have not already been developed and tested on animals, will be so before they are first applied to humans. For example, increased longevity or life extension is a prominent goal of human

enhancement. [...] The technology to achieve this in humans is yet to be realised, but startlingly effective advances in increasing longevity have already been made in animals' (Chan, 2009, p. 679).

12 Furthermore, in veterinary medicine, it is really difficult to find adequate funding for research on diseases affecting animals of little economic interest to the human being, including wild animals (Rathbone and Brayden, 2009).

13 For a discussion on tail docking, see the interesting paper by Fox, 2010.

14 See, for example, the statement of the *AVMA Animal Welfare Committee* about ear and tail docking in dogs at http://www.avma.org/issues/policy/animal_welfare/tail_docking.asp.

15 See the page of the American Kennel Club on the bulldog, http://www.akc.org/breeds/bulldog/.

16 After reproductive problems, mastitis is the second major cause of mass-killing in milk cows (see Ferrari et al., 2010).

17 The veterinarian Meyers-Wallen argues: '[I]nformation available from genetic research will only be useful in improving canine health if veterinarians have the knowledge and skills to use it ethically and responsibly. There is not only great potential to improve overall canine health through genetic selection, but also the potential to do harm if we fail to maintain genetic diversity. Our profession must be in a position to correctly advise clients on the applications of this information to individual dogs as well as to populations of dogs, particularly purebred dogs' (Meyers-Wallen, 2003, p. 73).

18 The Methusaleh Foundation, founded by the Transhumanist Aubrey de Grey, supports anti-aging research and holds, since 2003, the MPrize for the best transgenic mouse model with an extended lifespan. See http://www.mprize.org/.

19 This point cannot be further explored in this chapter. For a fruitful critique of the neoliberal logic in which the enhancement project is embedded, see Birch, 2008.

20 See, for example: http://www.centerforgreatapes.org/residents-details.aspx?id=44.

21 For the environmental effects of animal production, see the FAO Report by Steinfield et al., 2006.

22 Only in a footnote do Persson and Savulescu (2012) refer to the interests of non-human animals as also being at risk if we do not extend welfare concerns globally and into the remote future. However, they refuse to develop sustainable policies based on the moral status of animals, since this is contested (see p. 103).

References

Birch K. (2008) 'Neoliberalising bioethics: bias, enhancement and *economistic* ethics', *Genomics, Society and Policy*, 4(2), 1–10.

Bostrom N. and Savulescu J. (2009) 'Human enhancement ethics: the state of the debate', in Bostrom N. and Savulescu J. (eds), *Human Enhancement* (Oxford and New York: Oxford University Press), pp. 1–22.

Chan S. (2009) 'Should we enhance animals?', *Journal of Medical Ethics*, 35, 678–83.

Church S.L. (2006) 'Nuclear transfer saddles up', *Nature Biotechnology*, 24, 605–7.

Coenen C. (2010) 'Deliberating Visions: The Case of Human Enhancement in the Discourse on Nanotechnology and Convergence', in: Kaiser M., Kurath M., Maasen S. and Rehmann-Sutter C. (eds) *Governing Future Technologies: Nanotechnology and the Rise of an Assessment Regime* (Dordrecht: Springer), pp. 73–87.

Cooper M. (2008) *Life as Surplus: Biotechnology and Capitalism in the Neoliberal Era* (Washington: University of Washington Press).

Donovan D.M. (2005) 'Engineering disease resistant cattle', *Transgenic Research*, 14(5), 563–7.

Dvorsky G. (2006) 'All together now: developmental and ethical considerations for biologically uplifting nonhuman animals', *Journal of Personal Cyberconsciousness*, 1(4), http://web.archive.org/web/20070108172808/http://ieet.org/writings/AllTogetherNow.pdf.

European Parliament STOA (2009) *Human Enhancement Study*, http://www.europarl.europa.eu/RegData/etudes/etudes/join/2009/417483/IPOL-JOIN_ET(2009)417483_EN.pdf.

Ferrari A. (2006) 'Genetically modified laboratory animals in the name of the 3Rs?', *ALTEX*, 23(4), 294–307.

Ferrari A. (2008) *Genmaus & Co. Gentechnisch veränderte Tiere in der Biomedizin* (Erlangen: Harald Fischer Verlag).

Ferrari A. (2010) 'The control nano-freak: multifaceted strategies for taming nature', in Kjolberg K. and Wickson F. (eds), *Nano Meets Macro Social Perspectives on Nano-scaled Sciences & Technologies* (Singapore: Pan Stanford Publishing), pp. 307–35.

Ferrari A. (2012a) 'Animal enhancement: Künftiger Alptraum für Nutztiere?', http://www.tier-im-fokus.ch/nutztierhaltung/animal_enhancement/.

Ferrari A. (2012b) 'Animal disenhancement for animal welfare: the apparent philosophical conundrums and the real exploitation of animals. A response to Thompson and Palmer', *Nanoethics*, 6, 65–76.

Ferrari A. (2013) 'Zwischen Tierschutz und Ausbeutung: Animal Enhancement als Herrschaftsprojekt', in Rippe K.-P. and Thurnherr

U. (Hrsg.), *Tierisch Menschlich. Beiträge zur Tierphilosophie und Tierethik* (Erlangen: Harald Fischer), pp. 97–114.

Ferrari A., Coenen C., Grunwald A. and Sauter A. (2010) *Animal Enhancement. Neue technische Möglichkeiten und ethische Fragen* (Bern: Bundesamt für Bauten und Logistik BBL), http://www.ekah.admin.ch/fileadmin/ekah-dateien/dokumentation/publikationen/EKAH_Animal_Enhancement_Inh_web_V19822.pdf.

Forsberg C.W. et al. (2003) 'The Enviropig physiology, performance, and contribution to nutrient management, advances in a regulated environment: the leading edge of change in the pork industry', *Journal of Animal Science*, 81(14 Suppl. 2), E68–77.

Fox M. (2010) 'Taking dogs seriously?', *Law, Culture and the Humanities*, 6(1), 37–55.

Galli C. et al. (2003) 'Pregnancy: a cloned horse born to its dam twin', *Nature*, 424, 635.

Galli C. et al. (2008) 'Somatic cell nuclear transfer in horses', *Reproduction in Domestic Animals*, 43(Suppl. s2), 331–7.

Golovan S.P., Meidinger R.G., Ajakaiye A., Cottrill M., Wiederkehr M.Z., Barney D.J., Plante C., Pollard J.W., Fan M.Z., Hayes M.A., Laursen J., Hjorth J.P., Hacker R.R., Phillips J.P. and Forsberg C.W. (2001) 'Pigs expressing salivary phytase produce low-phosphorus manure', *Nat. Biotechnol.*, 19, 741–5.

Gottlieb S. and Wheeler M.B. (2008) *Genetically Engineered Animals and Public Health: Compelling Benefits for Health Care, Nutrition, the Environment, and Animal Welfare* (Washington DC: Biotechnology Industry Organization), http://www.bio.org/foodag/animals/ge_animal_benefits.pdf.

Gruen L. (2011) *Ethics and Animals: An Introduction* (Cambridge: Cambridge University Press).

Harris J. (2007) *Enhancing Evolution* (Princeton and Oxford: Princeton University Press).

Harvey D. (2007) *A Brief History of Neoliberalism* (Oxford: Oxford University Press).

Henschke A. (2012) 'Making sense of animal disenhancement', *Nanoethics*, 6(1), 41–6.

Hubrecht R. (1995) 'The welfare of dogs in human care', in Serpell J. (ed.), *The Domestic Dog: Its Evolution, Behaviour and Interventions with People* (Cambridge: Cambridge University Press), pp. 179–98.

Hughes J. (2004) *Citizen Cyborg. Why Democratic Societies Must Respond to the Redesigned Human of the Future* (Cambridge: Westview Press).
Kimmelmann B. (1983) 'The American Breeders' Association: Genetics and Eugenics in an Agricultural Context, 1903–1913', *Social Studies of Science*, 13, 163–204.
Kues W.A. and Niemann H. (2011) 'Advances in farm animal transgenesis', *Preventive Veterinary Medicine*, 102, 146–56.
Lebedev M.A. and Nicolelis M.A. (2006) 'Brain-machine interfaces: past, present and future', *Trends in Neuroscience*, 29(9), 536–46.
Lehrer J. (2009) 'Neuroscience: small, furry... and smart', *Nature*, 461(7266), 862–4.
Lenk C. (2002) *Therapie und Enhancement. Ziele und Grenzen der modernen Medizin* (Berlin: Springer).
Liao S.M., Sandberg A., Roache R. (2012) 'Human Engineering and Climate Change', *Ethics, Policy and Environment*, 15(2). 206–221. DOI: 10.1080/21550085.2012.685574.
Lush J.L. (1937) *Animal breeding plans* (Ames: Iowa State College Press).
Lyons L.A. (2010) 'Feline genetics: clinical applications and genetic testing', *Topics in Companion Animal Medicine*, 25(4), 203–12.
Macchiarini F. et al. (2005) 'Humanized mice: are we there yet?', *Journal of Experimental Medicine*, 21, 202(10), 1307–11.
Maga E.A., Shoemaker C.F., Rowe J.D. et al. (2006) 'Production and processing of milk from transgenic goats expressing human lysozyme in the mammary gland', *Journal of Dairy Science*, 89, 518–24.
Mamiya T., Yamada K., Miyamoto Y. et al. (2003) 'Neuronal mechanism of nociceptin-induced modulation of learning and memory: involvement of N-methyl-D-aspartate receptors', *Molecular Psychiatry*, 8(8), 752–65.
Meyers-Wallen V.N. (2003) 'Ethics and genetic selection in purebred dogs', *Reproduction of Domestic Animals*, 38, 73–6.
Palmer C. (2011) 'Animal disenhancement and the non-identity problem: a response to Thompson', *Nanoethics*, 5, 43–8.
Panarace M. et al. (2007) 'How healthy are clones and their progeny: 5 years of field experience', *Theriogenology*, 67, 142–51.
Pearce D. (2007) 'The abolitionist project. Text adapted from invited talks given at the Future of Humanity Institute (Oxford University) and the Charity International *Happiness Conference*', http://www.abolitionist.com/.

Pearce D. (2011) 'Transhumanism 2011. Interview with David Pearce', *Manniska Plus*, http://www.hedweb.com/transhumanism/overview2011.html.

Persson I., Savulescu J. (2012) *Unfit for the Future. The Need for Moral Enhancement.* Oxford: Oxford University Press.

President's Council on Bioethics (2003) *Beyond Therapy: Biotechnology and the Pursuit of Happiness* (Washington DC: The President's Council on Bioethics), http://bioethics.georgetown.edu/pcbe/reports/beyondtherapy/beyond_therapy_final_webcorrected.pdf.

Rathbone M. and Brayden D. (2009) 'Controlled release drug delivery in farmed animals: commercial challenges and academic opportunities', *Current Drug Delivery*, 6(4), 383–90.

Raven P.G. (2011) Uplift ethics and transhuman hubris, http://futurismic.com/2011/07/26/uplift-ethics-and-transhuman-hubris/?utm_source=feedburner&utm_medium=feed&utm_campaign=Feed%3A+futurismic_feed+%28Futurismic+-+the+fact+and+fiction+of+tomorrow%29.

Roco M. and Bainbridge W. (eds) (2002) *Converging Technologies for Improving Human Performance. Nanotechnology, Biotechnology, Information Technology and Cognitive Technology*, NSF/DOC-sponsored report (Arlington: World Technology Evaluation Center), http://wtec.org/ConvergingTechnologies/1/NBIC_report.pdf.

Rollin, B.E. (1995) *The Frankenstein Syndrome. Ethical and Social Issues in the Genetic Engineering of Animals.* Cambridge: Cambridge University Press.

Savulescu J. (2011) 'Genetically modified animals: should there be limits to engineering the animal kingdom?', in Beauchamp T. and Frey R. (eds), *The Oxford Handbook of Animal Ethics* (Oxford: Oxford University Press), pp. 641–70.

Schaffer M. (2009) *One Nation under Dog* (New York: Henry Holt & Co.).

Schultz-Bergin M. (2014) 'Making better sense of animal disenhancement: A reply to Henschke', *Nanoethics*, 8, 101–9.

Singer (2011) *Practical Ethics*, 3rd edn (Cambridge: Cambridge University Press).

Steinfield H. et al. (eds) (2006) *Livestock's Long Shadow: Environmental Issues and Options* (Rome: Food and Agriculture Organisation), http://www.fao.org/docrep/010/a0701e/a0701e00.htm.

Talwar S.K. et al. (2002) 'Behavioural neuroscience: rat navigation guided by remote control', *Nature*, 417, 37–8.

Tang Y., Shimizu E., Dube G.R. et al. (1999) 'Genetic enhancement of learning and memory in mice', *Nature*, 401(6748), 63–9.

Thompson P. (2008) 'The opposite of enhancement: nanotechnology and the blind chicken problem', *Nanoethics*, 2, 305–16.

Wall R.J., Powell A.M., Paape M.J. et al. (2005) 'Genetically enhanced cows resist intramammary Staphylococcus aureus infection', *Nature Biotechnology*, 23, 445–51.

Wolbring G. (2008) 'Why NBIC? Why human performance enhancement?', 21, *Innovation; The European Journal of Social Science Research*, 1, 25–40.

Wolbring G. (2009) 'Die Konvergenz der Governance von Wissenschaft und Technik mit der Governance des "Ableism"', *Technikfolgenabschätzung – Theorie und Praxis*, 2(18), 29–35.

Wolbring G. (2010) 'Human enhancement through the ableism lens', *Dilemata*, 3, http://www.dilemata.net/revista/index.php/dilemata/article/view/31/46.

Young L. (2009) 'Pet economy: meet the fur babies', *Telegraph.co.uk*, 5 November, http://www.telegraph.co.uk/health/petshealth/6507575/Pet-economy-meet-the-fur-babies.html.

2
Improving Animals, Improving Humans: Transpositions and Comparisons

Florence Burgat

Abstract: *Animal improvement and human improvement do not have the same status, nor do they have the same goal. Human improvement supposedly aims to increase the well-being and health of individuals by acting on their appearance, their mental or physical capacities, their emotional state, in ways the subject desires, with a view to what they consider to be a better life. The objective of animal improvement has never been to make the life and well-being of animals better; it is rather to improve the characteristics of animals that are useful to humans. These points are explored by retracing the main stages of animal improvement in France since the 18th century and its present aims as determined by the law on breeding of 28 December 1966.*

Bateman, Simone, Jean Gayon, Sylvie Allouche, Jérôme Goffette and Michela Marzano, eds. *Inquiring into Animal Enhancement: Model or Countermodel of Human Enhancement?* Basingstoke: Palgrave Macmillan, 2015.
DOI: 10.1057/9781137542472.0007

1 Position of the problem

When talking about improvement of humans and improvement of animals, does this imply an approach that is similar in spirit, aimed at the same goals and using the same techniques? Does one in both cases do good to individuals, improve their life and their health? Is it relevant to use the same term – improvement – when, as it will be seen, the objectives of improvement in each case are radically different?

In attempting to compare and transpose the improvement of animals to that of humans, I will start by saying that although these two areas of improvement share some common traits, they have, fundamentally, nothing to do with each other. I would also like to point out, right from the beginning, that there is a lack of debate if one considers that the only problem posed by improving animals is that of its possible transposition to humans. This point of view has many advocates, including the French philosopher François Dagognet (1988, p. 138) who accepts that 'the suffering of hens raised under ultraviolet rays or calves that are deliberately paralysed to fatten them' can only be condemned 'if these brutal and violent actions risk contaminating men and lead them to cruelty'. It is, in fact, thanks to the improvement of animals that these breeding methods are feasible.

The improvement of animals raises specific problems that should not be ignored. There are two types of problems: the first concerns the species, that is to say, the genetic impoverishment generated by the improvement of animals since this is aimed at creating, and not only retaining, standards that are in keeping with the market's expectations; while the second has to do with individuals, the damage caused to the health and well-being of most of the individuals concerned. This applies to livestock (bred for food consumption), laboratory animals and pets.

If we genuinely wish to conduct an exercise of comparison and transposition, it is necessary to know precisely what animal improvement entails. We need to know what techniques are applied to animals, how, since when and to what end. It then becomes quite clear that human improvement and animal improvement are quite distinct practices despite some common traits.

2 What is animal improvement?

What is improvement? According to various dictionaries, 'to improve' means to make something better, to increase the value, quality or productivity of something, to enhance or ameliorate, to increase profitability, excellence or desirability, or even to alter or add to an existing subject of invention or discovery something that does not destroy its identity or character but accomplishes greater efficiency. In fact, the Webster dictionary even specifies farm stock as an example.

In this particular case, making something or someone better is not axiologically neutral; one improves the subject for its own good or for its user. Similarly, one can improve something to embellish it or in the interest of the user. *One then improves according to an orientation that is not good in itself but for some ulterior motive.*

In fact, to illustrate the French word for improvement (*amélioration*), the Larousse dictionary quotes a phrase by the author Jean Giono: 'I propose to go into animal farming there and to improve the bovine breed' ('Je propose d'y faire de l'élevage et d'améliorer la race bovine'). Is the bovine breed improved for the well-being of cattle and in their interest? Certainly not! It is improved for the advantage of breeders, and it is in their interest to obtain more milk, eggs, wool, leather and meat from their animals. Zootechnics has successfully faced the challenge of improvement. For example, in France between 1961 and 1963, the number of eggs laid per hen rose from 200 to 210 a year to 285 in 1994, and 300 at present. Between 1950 and 1980, meat production, with the exception of poultry farming, grew from 1.96 to 3.4 million tons, while milk from cows increased from 150 to 312 million hectolitres. The growth period of hens bred for their meat decreased in 30 years (1960–90) from 80 days to 30 days (Jussiau et al., 1999, pp. 440–2). Canadian, American and Singaporean geneticists have succeeded in producing salmon with a weight eleven times higher than normal, while Israeli researchers have created a featherless chicken, the Totally Naked Rooster, in order to save money on ventilating coops for the breeders and on plucking feathers for the retailers.

What is animal improvement? What does one improve precisely, and why? Right from the start, animal improvement only had one objective: to select the characteristics and develop those performances that are economically useful to man. It is indeed the *improvement* of animals that is discussed in scientific articles written by biologists or zootechnicians. In fact, in France, the notion was fixed and institutionalised by legislation passed in 1966 on breeding, 'the purpose of which is to improve the quality and conditions of the livestock of cattle, pigs, sheep and goats'. The geneticist François Minvielle (1998), however, uses the more neutral and strictly descriptive term of 'animal selection'.

Improving animals is part of the process of their domestication. But historians of animal breeding agree that the origin of improvement, in the sense of a coordinated and systematic undertaking, goes back to the late 18th century, when the industrialisation movement disrupted all production modes. The farm animal was no longer an auxiliary of cereal production, but became a productive animal at the heart of a mixed farming/breeding system. A new science emerged: zootechnics. It became an academic discipline as of 1894. Its slogan was clearly stated in textbooks. The French naturalist Eugène Baudement (1816–63) defined the zootechnician as an 'engineer of living machines'. He added that 'animals are living machines not in the figurative sense of the word but in its most rigorous acceptance [...]. The better we know the construction of these machines, the laws of their functioning, their requirements and resources, the better we can exploit them safely and to our advantage' (in Jussiau et al., 1999, p. 362). The improvement of cattle lies at the core of this new science. It explores cross-breeding, on the one hand, and selection of pure races through the best breeding stock, on the other. The first artificial insemination was tried out on a dog by an Italian scholar named Lazzaro Spallanzani (1729–90). But it was not until the 19th century that new trials were carried out, particularly in Russia. The biotechnologies of reproduction (artificial insemination, cryopreservation of embryos and so on) have since become an important sector in animal improvement. All species of animals reared for food consumption are concerned, as well as horses.

The trend towards improvement started out in England. Although the attempt to 'improve' bovine breeds for the purpose of meat production by crossing them with the English Durham breed failed, it marked the beginning of a process and introduced the notion of 'improved cattle'. Horses for reproductive purposes were imported from England by the Haras du Pin, a stud farm created in 1715 by order of the King of France, Louis XIV, the construction of which was completed in 1730. Currently owned by the French State, it continues its genetic research, manages the equine data files and houses stud stallions. Each breed is associated with a region. Each race has common characteristics in terms of morphology and behaviour that are fixed empirically by the breeders. A 'standard' is thus established. The animals complying with this standard are noted down in a genealogical register to record their pedigree. In the 1930s, the Ministry of Agriculture registered a certain number of breeds. These registers were the basis for the organisation of breeding until 1960.

38   *Florence Burgat*

There was subsequently a gradual elimination of animals without clearly affirmed features. The situation became drastic. The diversity of races shrank, and a few 'conservatories of breeds' are now sheltering animals of the past, in case we should need their services one day.

The theory of perfection developed by Eugène Baudement demonstrates how animal improvement, far from contributing to the development of the qualities of each individual for its own interest, reduces it to a single function, as far as it is possible. According to him, 'Perfection is a set of all the characters that best respond to the destination of an animal; it is all the combined qualities that, to the exclusion of all the others, make the animal capable of one single type of service [...], in other words, to constitute a machine of maximum output' (in Jussiau et al., 1999, p. 362).

The genetic improvement of cattle, based on a rediscovery of Mendel's laws (1907), has become one of the major focuses of zootechnics, next to conducting research on food, the environment and reproduction. *Animal Breeding Plans* (1937), written by the biologist Jay Laurence Lush, is a pioneering work in the field of genetic improvement of production animals, in which he puts forward a general theory on this topic. This work is at the origin of the first genuine genetic improvement programmes. Genetics is making constant progress: from the quantitative genetics of the 1960s to the most current tools of post-genomics, still known as functional genetics, we must add the technologies of molecular and bio-informatic analyses.

Genetic improvement, for which the French law on breeding of 28 December 1966 continues to provide the framework, has become institutionalised, notably with the creation in 1946 of the National Institute for Agricultural Research (*Institut National de la Recherche Agronomique*, INRA), and that of a specific research section within the Institute on 'Animal Production' in 1948. François Tanguy-Prigent, the Minister of Agriculture at the time, defined the missions of the Institute in the following terms: 'To organise, execute and publish all research work on the improvement and development of animal and plant production, as well as the preservation and transformation of agricultural products' (in Jussiau et al., 1999, p. 419). Furthermore, the National Commission for Genetic Improvement (*Commission Nationale pour l'Amélioration Génétique*, CNAG) is responsible for submitting proposals to the Ministry of Agriculture and Fishing on the methods and means for improving the bovine, porcine and ovine stocks. The CNAG collaborates with several research and technical institutes (*Institut Technique du Porc*, 1961; *Institut Technique de l'Élevage Bovin*, 1962; *Institut Technique des Ovins et des Caprins* (ITOVIC), 1967; *Institut Technique de l'Aviculture*, 1968; and so on), as well as with centres involved in seed production, insemination and 'racial databases', all under the aegis of the Ministry of Agriculture.

It should be noted that zootechnics does not have any branch devoted to animal behaviour. Its action is aimed at adapting the animal to its environment. Jean-Michel Faure and Andrew Mills (1995) point out that there are two options: one either adapts the environment to the ethological needs of animals, or one adapts animals to the industrial environment. The latter option was chosen. It 'consists in taking action on the animal' (Faure and Mills, 1995). Several types of intervention are carried out: surgery, administering medication, interventions at certain stages of development and modification of genetic identity. 'Veterinary interventions, such as debeaking, dehorning, injection of tranquilisers […] are also looked down on by protectionists (sic) and consumers, and their use tends to be restricted by law. We are therefore only interested in the methods designed to improve well-being through ontogenetic or genetic means' (Faure and Mills, 1995). Further on, they add:

> It is nevertheless increasingly important to select domestic animals, and especially birds, on the basis of their behaviour because their environment of selection is different from that of their breeding. This is particularly true of hens bred for their meat or for their descendants, [the first of] which are always bred on the ground while parental layers are selected in cage batteries. Until the middle of the 1960s, 'hens bred for their meat' were selected on the ground, while hens with a behaviour pattern of laying eggs on the ground (outside trap nests) or brooding were eliminated. Ever since selection has been carried out in cages, animals likely to display this pattern of behaviour are no longer eliminated because the selection environment does not permit it, and one is increasingly coming across the problem of egg laying on the ground or brooding among meat breeding fowl.[1] (Faure and Mills, 1995)

Until recently, the well-being of animals was evaluated exclusively on the basis of zootechnical criteria, productivity being the indicator of well-being. Although certain biologists are engaged in a more global reflection on well-being, the prevailing trend is still focused on 'genetics only', linked to the inevitably reductionist context of improvement. In consequence, the studies on animal well-being conducted by the French National Institute of Agricultural Research (INRA) have incorporated a new network, 'Genetics of animal adaptation and well-being' (Beaumont et al., 2005), in which well-being is understood as 'an adaptation to the environment'.[2] As it has just been seen, the choice of adapting animals to the environment is not a new one, but no thought had previously been given to the idea of including it in a programme on 'animal well-being'.

3 The first limitations of the comparison

Although human enhancement is willingly described by its advocates as the attainment of a state that is 'better than well', the aim of animal improvement has never been to offer them such a state. In fact, the opposite is true, and in an increasingly obvious way, due to the growing capacities for selection applied to livestock,[3] but also to pedigree cats and dogs, given that the hyper-specialisation of pedigrees, including the miniaturisation of animals, leads to genetic pathologies.[4] Two examples illustrate this. The first one concerns dogs.[5] Pedigrees are systematically sought by people who want to acquire 'pets', but there are also breeds that are made fashionable by various agents who, in fact, promote a product. Presenting animals with highly distinctive characteristics in different types of media creates a demand for dogs identical to the exhibited and promoted hyper-type. To satisfy these client demands – that are generated from start to finish –, breeders 'produce' hyper-inbred animals that are particularly sensitive to certain diseases and have a shorter lifespan. Hyper-types are animals with an accentuated morphology of the pedigree in question and features that have been deliberately exaggerated. Today, the trend for hyper-types is notable, and it is encouraged by pedigree clubs, competitions and their judges, breeders and so on. The other side of the coin is that some pedigrees disappear because the function they used to carry out is outdated (certain types of sheep dog) or, less rapidly, because they are no longer 'in fashion'.

The second example concerns cattle bred for meat. The 'double muscling gene' provokes an excessive muscular development of the hindquarters. The Belgian Blue Breed is an illustration. These animals are so horrendous that the females cannot give birth without a caesarean. It has also been observed that they suffer from weak front legs because of overweight. The latter problem generates specific pathologies (limping, split pads) and is common to nearly all reared animals that have increasingly been made bigger and therefore too heavy for their skeleton, as a result of research on food and genetic improvement. The outcome is painful pathologies, physical handicaps and behavioural problems, but these, admittedly, do not undermine the quality of the meat.

Animal improvement began using empirical selection with a view to emphasising certain features. In the case of dogs as well as production animals, 'standards' were set and recorded in genealogical registers during the second half of the 19th century and the beginning of the 20th century. Improvement was subsequently based on the increasingly advanced tools of genetics. Its aim was to multiply productivity and performance by ten (yield in eggs, milk and meat, and capacity for racing) or to modify phenotype (pets). The goal of reproductive biotechnologies is to develop the 'descendants of elite sires'[6] (through artificial insemination, cryopreservation of embryos, embryo transfer), as part of the bovine and porcine livestock improvement programme. Cloning and transgenesis are biotechnologies with applications that are mainly agronomic (Jussiau et al., 2006, pp. 67–9).[7]

We can thus make the following observations:

A. First, animal improvement and human improvement *do not have the same status at all*: the former has existed for a long time *as such* and has since become institutionalised; its ultimate goals are defined by law and are handled by technical and research institutes. As for the second – human improvement – it clearly has totally distinct goals from those of animal improvement. It is an area that is gradually emerging, accompanied by numerous problems, as a result of a supply and demand that has been made possible by techniques proposing 'something more' or 'something better'. Philosophers, sociologists and physicians question the justification of these demands, even when they come from the interested parties themselves, concerning the nature of this 'something more' or 'something better'.

B. These two areas nonetheless share a possible *technical continuity*, in so far as the techniques that are being perfected for interventions in humans were previously developed on and applied to animals. This continuity is required for the extrapolation of human enhancement practices from those of animal improvement. This technical continuity is not specific to enhancement as it is the basis for animal experimentation preceding any kind of human experimentation.

There might, of course, be cause for alarm about animal improvement if one imagines that part of humanity will have the technical resources and political power to subjugate another part of humanity by shaping it to serve its own interests and without any other concern whatsoever, exactly in the same way as humanity does with animals. And one is fully aware of all that Nazi eugenics owes to zootechnics.[8] This probably explains the insistence on distinguishing 'human improvement' from the 'old eugenics' by stressing the aspect of subjects being at liberty to decide for themselves what they feel is in their interest. It is up to other scholars to deal with the solidity of this opposition between a decision imposed by one part of humanity on another (bad eugenics) and a social atmosphere that 'freely pushes' everyone towards improvement, in a sense that is nevertheless predetermined. In any event, it would be hasty to associate, without further ado, the fiction of a part of humanity reduced to the consumable use of another with the reality of practices currently being applied to animals.

4 Relevance of transpositions and comparisons

A. It has been noted, on the one hand, that a number of practices may appear to be similar (and how could it not be so, since animal species constitute the testing ground for all that is or could be applied to human medicine?): improvement of a given physical feature, development of a particular capacity, acceleration of the growth rate of production animals *versus* an extension of the life of humans, alteration of their emotional state and so on. An alteration in the emotional state of birds, living six together in batteries of very narrow cages, has just been mentioned. If humans resort to chemistry for this,[9] animals are genetically modified in order to be better 'adapted' to industrial breeding conditions. Concentration of animals in a cage or building imposed by industrial breeding generates all kinds of pathologies and behavioural problems. To reduce the stress of pigs, the stress gene has been almost eliminated among pigs in Landrace. Birds (hens, quail) have been selected according to their lowest level of emotionalism. It is more in the interest of industrial breeding than that of animals to obtain apathetic individuals. In consequence, the phrase 'well-being and genetic adaptation of animals' appears to be problematic.

B. On the other hand, the reasons motivating interventions on animals and on humans are radically opposed.
   1. The goal of human improvement is to increase the well-being and health of individuals by providing a state that is 'better than well'. This means satisfying a desire concerning their appearance, acting on their mental or physical capacities, on their emotional state, *specifically in the way the subject wants it*, with a view to what they consider to be a better life.
   2. The objective of animal improvement, on the contrary, was never to make the life and well-being of animals better. It is not concerned with their well-being. In fact, these improvements are carried out even to their detriment. In any event, they are undertaken without any concern for the consequences on their well-being and health, provided, of course, that the damage resulting from this improvement does not undermine its objectives. Only the characteristics that are considered useful to humans are improved in animals. The damage caused to the animals is not taken into account for two kinds of reasons:
      ▶ First of all because animals are traditionally excluded from the field of ethical interrogation. Admittedly, for some time now, a number of ethicists have questioned the negative effects of such 'improvement' on the integrity of animals. They do not restrict themselves to arguments about 'interfering with nature' or 'playing God', but also concentrate on the damage caused in terms of the health and well-being of animals. In the circles of zootechnics and breeding, ethical questions are centred on 'social acceptability'. In other words, in their view, what is ethical is what society accepts, hence the considerable investments made in communicating on their research in an effort to ensure that it is acceptable to public opinion. The inflated vocabulary of ethics is notable in this communication effort, especially in the areas of experimentation, industrial breeding, transport, slaughter and hunting. By talking constantly of 'ethics', 'respect' and 'well-being', the debate on the justification of these practices is drowned or anaesthetised. In reality, the purpose of these – inappropriate – terms is to serve as a reminder of the most basic rules of professional ethics that 'unnecessary suffering' to animals should be avoided.[10]

▶ Secondly, the damage caused to animals by improvement is not taken into consideration for purely economic reasons. Since the lifespan of animals bred for commercial purposes or laboratory experiments is limited to the period of fattening, the task to be accomplished or the response to be provided, the conditions in which they live for a few weeks or months is of no importance. For example, the paws of rabbits reared in a cage with a wire-netting floor are split because of their weight: they have been made heavier and bigger by genetic means, that is to say, they are too heavy and big. As long as the damage does not hamper production, it is hardly ever taken into account because this would entail changing the entire breeding system.

Let us continue this comparative exercise. In the case of humans, individual freedom seems to be exalted by decisions to improve oneself. The subjects decide, in principle with full knowledge of the issue at stake, to increase their performances in a given area. In this way, they are the sole masters of their life and judge of the risks they are willing to take in order to fulfil a goal to which they have deliberately given priority. This, at least, is the feeling they have. Obviously, it is not at the levels of enlightened decision-making, of autonomy and of self-determination that we seek to justify comparisons between human improvement and animal improvement.

Whereas some people plan to improve the entire human species, others are worried about the criteria and, therefore, the aims that are implied by this improvement. They ask themselves whether a new figure of humanity subsumed to purposes other than the interests of each individual might not be emerging, or re-emerging. It is clear that genetic selection techniques, well controlled among mammals, can be extrapolated to humans. Some of them have already been applied for therapeutic purposes. As an extension of the Nuremberg Code (1947), and following its spirit, ethics committees assess the compatibility of these applications with a certain number of values recognised in the human being in an intrinsic manner.

5 Excursus: 'Genetics and selection: a distinction must be made between humans and animals'

In this regard, it is difficult not to smile at the embarrassment of three authors, breeding historians and zootechnicians who, in order to dissipate misunderstandings, felt they had to add to their book a lengthy special section entitled 'Genetics and selection: a distinction must be made between humans and animals' (Jussiau et al., 1999, pp. 378–81). And what a series of redundant attempts to state that: no, humans and animals are not the same! It is amusing to note the reappearance in their discourse of a metaphysical principle – chance – used here as a way of setting humans apart from a genetic research programme that aims to eradicate all chance occurrences in animal reproduction – a programme which is nonetheless based on characteristics common to all members of the animal species, to which humans belong. It is precisely this biological continuity between human and non-human animals that makes them uneasy. 'Contrary to common parlance [...] the human species, like all superior animals, procreates and does not reproduce'. Unpredictability, lottery or chance are all at the core of this phenomenon. One can legitimately ask who the 'superior animals' are here, if they are not mammals, therefore most of the animals subjected to selection. The confusion is great among our authors since they tend to mix different levels in their arguments. We are inclined to warn them not to apply to humans the principles that are applied to domestic animals because, as they themselves argue, all human beings are born free and equal!

Although they assure us that it would be criminal to do to humans what is done to animals, it is never a question of wondering why what is done to animals is not criminal. There are things, they say, that should only be applied to animals. We are never told why, but the matter is immediately put forward as if it were necessary to exonerate zootechnicians in advance from being accused of having done everything to allow criminal temptations and doubtful extrapolations. Thus, 'better', 'superiority', 'inferiority' and 'hierarchy' are good concepts to describe what is done to animals but criminal if applied to humans. 'Beware of wandering off the path', it is written! We are reminded of all that the particularisation of breeds owes to human creation, in order to state clearly that what must be preserved in humans is their diversity. And in conclusion, they once again thank the members of the French Constituent Assembly of 1789, who had the good idea of declaring that 'men are born free and equal'.

6 Conclusion

Today, thanks to tools combining genetics and molecular biology, we are capable of shaping organisms, and adjusting them to agronomic or scientific requirements. Let us recapitulate: we can increase the size of animals, their rate of growth and their productivity; we can obtain transgenic animals: (a) for agronomic purposes, for example, a change in the composition of certain processed foodstuffs such as milk, by diminishing the level of lactose that is indigestible for adults, and by increasing the quantity of the most sought after caseins for cheese production; and (b) for medical purposes, by obtaining models capable of developing human diseases or of supplying transplants, and by obtaining 'bio-reactors' capable of providing in their milk (mice, rabbits, goats etc.), and perhaps in their eggs, recombinant proteins for pharmaceutical, and even industrial uses (Darryl and Macer, 1990, pp. 112–13; Houdebine, 1998, pp. 157–70). Transgenic animals are organisms that have a genome in which one or several genes from another species have been introduced. The introduction of human transgenesis makes it possible to bring the animal biologically closer to the human. What best distinguishes human improvement and animal improvement is the fact that the former claims to serve the interests of the subject whereas the latter never serves the interest of the subject but derives benefit from it.

From the empirical selection of a given characteristic to genetic improvement, the differences are not only quantitative. A qualitative step has been taken since production is no longer dependent on the capacities inherent in the species. These advances make it possible to shape or produce animals according to determined needs. The performances of an organism are no longer simply optimised, they are also modified. Cloning and transgenesis, in particular, have freed the reproduction of living entities from several biological constraints that were once thought impossible to overcome. The logic has been reversed thanks to a technical mastery that makes it possible to adapt animals to the imperatives of production and not the other way round. So much so that one can ask whether we have not entered the era of 'trans-animalism'. This is a theory that we can perhaps test in the future.

Notes

1 I am the one who emphasises that all the studies on adapting animals to production systems were renamed 'well-being'.
2 For a presentation of the institutional and epistemological context underlying the definition of well-being as an adaptation to the environment, see Burgat and Dantzer (2001).

3   'Is it at all acceptable, ethically speaking, to alter the genome of animals according to the needs and desires of farmers and the market?' (Gamborg et al., 2005, p. 14). Some people feel that this is only acceptable if the goal is to improve the well-being and health of animals. But can one incorporate the genetic adaptation of animals to industrial breeding systems in this section?

4   The reference manual used by French veterinarians is *Prédispositions raciales et maladies héréditaires du chien et du chat* (Gough and Thomas, 2009).

5   My remarks are based on 'Réflexions sur la génétique du chien de race et le nécessaire maintien d'une variabilité génétique', article resulting from the Seminar of the *Société Française de Cynotechnie* (Denis, no date).

6   'Le taux de mise à bas obtenu après insémination sur un cycle est de l'ordre du 90%.' [The calving rate obtained after one cycle of insemination is about 90%.] (Mermillod et al., 2003, p. 324).

7   See also: Murray et al. (1999); Houdebine (2001, pp. 119–46).

8   François Dagognet (1988, p. 137), in a chapter that is both apologetic and deliberately horrific on industrial breeding and the biotechnologies underlying them, refers to ' "concentrational breeding" about which one justifiably believes it inspired the "death camps"'. It is also useful to quote, from a different angle, the American historian Charles Patterson, *Eternal Treblinka* (2002, pp. 125–62), and the edifying chapter on 'Improving the herds – from animal reproduction to genocide'.

9   See the chapter by Ruud ter Meulen in the companion volume: Bateman et al. (eds) (2015).

10 I examined this point in 'Expérimentation animale: "un mal nécessaire"' (2009).

References

Beaumont C., Boissy A. et al. (2005) 'Réseau "Génétique de l'adaptation et bien-être animal"', *Bulletin de l'Académie Vétérinaire de France*, 158(3), 257–62.

Burgat F. (2009) 'Expérimentation animale: "un mal nécessaire"', *Revue Semestrielle de Droit Animalier*, 1, 193–201.

Burgat F. and Dantzer R. (eds) (2001) *Les animaux d'élevage ont-ils droit au bien-être?* (Versailles: INRA Publications).

Dagognet F. (1988) *La maîtrise du vivant* (Paris: Hachette).

Darryl R. and Macer J. (1990) *Shaping Genes. Ethics, Law and Science of Using Genetic Technology in Medicine and Agriculture* (New Zealand: Eubios Ethics Institute).

Denis B. (no date) 'Réflexions sur la génétique du chien de race et le nécessaire maintien d'une variabilité génétique' (Belley: Société Française de Cynotechnie), http://pronaturafrance.free.fr/chien3.html.
Faure J.M. and Mills A. (1995) 'Bien-être et comportement chez les oiseaux domestiques', *INRA Productions Animales*, 8(1), 57-67.
Gamborg C., Olsson A. and Sandoe P. (2005) *Farm Animal Breeding Related Ethical Concerns* (Copenhagen: Danish Centre for Bioethics and Risk Assessment).
Gough A. and Thomas A. (2009) *Prédispositions raciales et maladies héréditaires du chien et du chat* (Paris: Med'Com).
Houdebine L.M. (1998) *Les animaux transgéniques* (Paris: Technique & Documentation Lavoisier).
Houdebine L.M. (2001) *Transgenèse animale et clonage* (Paris: Dunod).
Jussiau R., Montméas L. and Papet A. (2006) *Amélioration génétique des animaux d'élevage. Bases scientifiques, sélection et croisements* (Dijon: Éducagri).
Jussiau R., Montméas L. and Parot J.C. (1999) *L'élevage en France. 1000 ans d'histoire* (Dijon: Éducagri).
Lush J.L. (1937) *Animal Breeding Plans* (Ames: Iowa State College Press).
Mermillod P. et al. (2003) 'Biotechnologies de la reproduction porcine', *Journées de la Recherche Porcine*, 35, 323-38.
Minvielle F. (1998) *La sélection animale* (Paris: PUF 'Que sais-je?').
Murray J.D., Anderson G.B., Oberbauer A.M. and McGoughlin M.M. (eds) (1999) *Transgenic Animals in Agriculture* (New York: CABI).
Patterson C. (2002) *Eternal Treblinka* (New York: Lantern Books).
ter Meulen R. (2015) "The moral ambiguity of human enhancement", in Bateman S., Gayon J., Allouche S., Goffette J. and Marzano M. (eds), *Inquiring into Human Enhancement: Interdisciplinary and International Perspectives* (Basingstoke: Palgrave Macmillan), pp. 86-99.

3
Harming Some to Enhance Others

Gary Comstock

Abstract: *Generally speaking, we modify animals' genomes to give their progeny traits that will indirectly improve human life. So-called intentional genetic 'enhancements' of animals, then, usually make the target animals worse-off. What rules should govern animal experimentation in which we harm some directly to enhance others indirectly? I criticize the abolitionist conclusions of animal rightists that all animal enhancements should be banned, and I criticize the permissive conclusions of speciesists that all such procedures should be allowed. I argue that current animal welfare law provides a defensible platform on which to begin building ethically justifiable policy in this area.*

Bateman, Simone, Jean Gayon, Sylvie Allouche, Jérôme Goffette and Michela Marzano, eds. *Inquiring into Animal Enhancement: Model or Countermodel of Human Enhancement?* Basingstoke: Palgrave Macmillan, 2015.
DOI: 10.1057/9781137542472.0008.

The Finnish cross-country skier Eero Mäntyranta (1937–2013) won seven Olympic medals, perhaps in part because he inherited a genetic condition called primary familial and congenital polycythemia. The condition is caused by a mutation in the erythropoietin receptor gene and may convey physiological advantages. In Mäntyranta's case, his blood carried greater densities of red cells, and his lung capacity was 50% greater than that of typically developing adults. We might have considered Mäntyranta's condition a 'genetic enhancement' had he acquired it as the result of scientists injecting the gene mutation into his parents' gametes. But, as it was, he inherited the condition naturally. Suppose, contrary to fact, that his mother and father were keen to give their child every competitive edge and requested that their doctor insert the new gene into their child's genome. Current laws in most countries would forbid the doctor from acting on the request. And yet governmental regulatory bodies are likely to face increasing pressure to allow the use of such technologies.

Policy makers will be loath to approve germline changes until the techniques have been proven to be safe and effective in so-called animal models. The animal experiments will entail risks to the animals. Which rules should govern proposals to put animals at risk of pain and suffering when the research goal is to perfect techniques for human enhancement? I address the issue here by criticising the abolitionist conclusions of many animal rightists (that *no* such experiments should be permitted) and the permissive conclusions of many speciesists (that *all* such experiments are permissible), and I argue that current animal welfare laws provide a workable platform from which to begin to form defensible policies in this area. In Section I, I define key terms; Section II proposes six lessons for human enhancement drawn from the methods and practices of animal breeding. The next two sections, III and IV, take up the question 'When if ever are we justified in harming research animals?' In Section V, I briefly outline two-level utilitarianism, and in Section VI identify two key policy changes that this theory requires in current animal welfare policy.

1 Definitions

The deliberate modification of an individual's genome to improve it or its progeny is intentional genetic enhancement, the conscious attempt to control the expression of traits. Changes in traits occur regularly across human generations without specific intervention as Mäntyranta's case exemplifies. The variation that results from ordinary evolutionary processes of random mutation, blind environmental filtration and drift is not at issue here.

Human enhancement research aims either to change a patient's somatic cells and so improve the individual's performance on some trait, or to change an individual's sex cells and so improve the performance of future generations, or both. Programmes of the second sort constitute a small subset of human genetic research for several reasons. It is easier to acquire funding to develop tools to diagnose and replace or alter defective genes known to be responsible for causing an individual's disease.[1] As public opinion is largely opposed to sex cell changes, it is more difficult and more controversial to acquire funding for research meant to cause potentially irreversible changes to future human generations. However, as scientists come to understand more thoroughly cellular and molecular processes, the public is likely to become more accepting of germline biotechnology. And it is incumbent on policy makers, therefore, to have a reasoned response to the question 'Which procedures, if any, may we test on animals?'.

It is, to begin, far from apparent that we are justified in harming animals to gain knowledge essential to benefit humans with crippling diseases such as Parkinson's. Such justifications almost always are couched in utilitarian terms: the benefits to humans of the research far outweigh the costs to the experimental animals. But if such claims are difficult to establish in 'easy' cases in which the protection of human health is involved, it is *prima facie* much more difficult to justify harming animals when the purpose is not restoring human function but improving it. For how can we justify causing an animal to suffer when our goal is, say, to add two inches to a professional athlete's vertical leap, or to help a batter to hit a ball harder? It is one thing to use animals to try to help disabled humans to walk or think clearly or escape from intrusive thoughts. It is something else entirely to exploit an animal to try to raise someone's performance from standard to peak levels.

My subject, in sum, is the deliberate modification of human somatic cells to improve the recipient organism's or its progenies' performance on some physical or psychological trait. We have engaged in such enhancement for centuries whenever farmers bred this cow with that bull to optimise the amount of, say, milk that the next cow would produce, or the amount of lean meat the next steer's carcass would contain. Such enhancements, therefore, are enhancements not for animals, but for us – the consumers of milk, meat, leather or other animal by-products.

In the second half of the 20th century, the discovery of DNA led to powerful new technologies, including cloning and xenografting, that have allowed researchers to produce new animal enhancements: sheep capable of forming human blood-clotting pharmaceuticals in their milk, mice that live 25% longer than controls, and rats that run mazes more quickly than their kin (Lee et al., 2004). Such results give new impetus to the questions about what we owe experimental animals and whether we have in place adequate rules to govern the way we treat them. They also show that the idea of enhancing an animal is ambiguous between improving an animal's trait in a way that is good for the animal and improving it in a way that is good for its owner.

I will return to this issue. I begin, however, by taking up a preliminary question, one posed to me by the editors of this volume. What can we learn from the history of animal breeding as we begin the era of human enhancement? At best an informal student of the history of farmers' attempts to change the physical and psychological traits of their animals, I hesitate to pose as an authority capable of stating 'lessons'. With this qualification of my credentials in mind, I hazard six themes that seem prominent in the literature.

2 Six lessons for human enhancement from enhancing animals

Expect the unexpected

Elsewhere I have suggested that the history of our relationship with so-called food animals falls into three oversimplified time periods: hunting, husbandry and science (Comstock, 2000a). Nomadic hunters gathered and ate whatever wild plants and animals were available, but in the Neolithic period subsistence pastoralists began to tame and confine animals, carefully choosing wild goats and sheep for features such as docility, high milk production or thick coat. The means of selection operated at a gross visual level but they were effective. Within a few thousand years, domesticated animals including dogs, cattle and pigs were meeting human needs for food, fibre, companionship and traction.

The pre-agricultural hunting of animals changed the physical and psychological traits of target animals but with only gradual effects that were unintentional and took hold only over many generations. Practitioners of husbandry, on the other hand, were deliberate and thoughtful about the traits they wanted to see, and the changes they introduced were immediate. Changes had to be observable in the next generation or they would not be selected for.

Sometimes the changes not only improved the lives of humans but were also good for the animals themselves. Hardier breeds often meant individuals with healthier lifespans resulting from reduced vulnerability to, for example, parasites. And yet husbandry also produced animals with changes that were good for humans and bad for animals. Pure bred dogs, such as German Shepherds, have a genetic propensity for hip dysplasia (Zhou et al., 2010). High producing cows are prone to higher incidence of mastitis. Belgian bulls, with muscles layered on other muscles resulting from natural knockouts of a protein gene, produce fetuses so large that caesarean delivery is a necessary part of maternal care (McPherron and Lee, 1997). Powerful Thoroughbred horses with increased muscle mass and lower bone density are put at risk of breaking fragile legs. Bred to fit into a purse, tiny dogs are so chronically anxious and subject to obsessive-compulsive behaviours and panic attacks that one might wonder whether their lives are worth living from their perspective (Overall et al., 2006; Overall and Dunham, 2002; Salgirli and Dodurka, 2011).

Animal science – the last and current epoch – has, as I say, brought an increasing power and precision to animal breeding. Researchers now directly manipulate genes, transferring them across species boundaries unbridgeable by sexual means of reproduction. Thus, for example, we have Atlantic salmon enhanced with antifreeze genes that grow to market weight much more quickly than their unenhanced relatives (Aqua Bounty Technologies Inc., 2010), and Enviropigs, transgenic hogs that can chemically degrade phytate – normally indigestible – and thereby reduce the phosphorus content of their manure by as much as 75% (Streiffer and Ortiz, 2010). The new enhancement technologies have made it possible to make animals' lives miserable, too. Cows injected with recombinant bovine growth hormone burn out faster than normals and must be culled from the herd earlier (Government of Canada, 2002). Hogs with human growth hormone genes produced in Beltsville, Maryland, were so arthritic and sickly that they had to be euthanized (Pursel et al., 1989, *cf.* Comstock, 2000b). Engineers purposely produce experimental rats with propensities to grow malignant tumours, or to shiver and tremble from multiple sclerosis symptoms induced by genetically caused deficiencies in their myelin proteins, or to suffer Elephant Man like skin conditions. Others are morbidly obese or aberrantly lack hair, a covering necessary to regulate basic functions and bodily homeostasis. And so on (Comstock, 2000b).

What will we see in humans? Non-genetically based physical enhancements have already produced, for example, the carbon fibre prosthetic legs of the runner Oscar Pistorius. This is only the beginning. The transhumanist Nick Bostrom (2003) speculates that intentional human enhancement 'will lead to more love and parental dedication' because '[s]ome mothers and fathers might find it easier to love a child who, thanks to enhancements, is bright, beautiful, healthy, and happy'. Whatever you make of this statement, the history of animal breeding suggests that you ought to expect the unexpected – both for better and for worse.

Expect the unexpected to arouse opposition

The first wave of agricultural biotechnologies provoked opposition. For example, recombinant bovine growth hormone and genetically engineered herbicide resistant crops became focal points for the ire of groups objecting to the technologies on ethical, social, environmental and religious grounds. As of November 2010, the Eurobarometer was reporting that 70% of Europeans believe that genetically modified (GM) food is unnatural. Over 50% believe it is not safe for them or for future generations (Directorate-General for Communication, 2010). Why do so many apparently oppose animal biotechnology? There are at least 20 ways to argue against agricultural biotechnology (Comstock, 2000b). They include objections on the grounds of the consequences to human health; to family farmers; to subsistence farmers; to scientists and to future generations; to ecosystems and plant germplasm diversity; and to research animals. They include principled arguments that claim that to engage in agricultural biotechnology is illegitimately to: play God; use nonsexual means to reproduce; arrogate historically unprecedented power to ourselves; exhibit arrogance and *hubris*; commodify life; and disrupt the integrity of creation. A popular and influential group of arguments holds that agricultural biotechnology is unethical because it is unnatural, which variously means to transfer the essence of one living being into another, or to change the *telos*, or end, of an individual.

I have argued that none of these arguments is sound. Some of them are so intuitively attractive, however, that many people are willing to accept them and mobilise groups to regulate or ban the technologies of genetic engineering. This leads us to the next point.

Expect opposition to fade eventually if someone is making money from the novelty

While the majority of Europeans continue to oppose GM crops and animals, attitudes differ in other parts of the world. Why have Monsanto and other multi-national corporations continued to produce GM products? Because large numbers of farmers outside Europe buy GM seeds and chemicals, and larger numbers of consumers in places such as the United States, Brazil and China have no problems eating the products. Farmers would quickly stop buying GM seeds if they were unable to sell their goods. But as long as GM products are thought to be safe by the governmental regulatory bodies of large consumer markets, and as long as consumers accept the judgments of those bodies that GM products are equally as safe and nutritious as non-GM competitors, those consumers will continue to buy GM products – especially if the GM lines are cheaper than the non-GM products.

Human genetic enhancement is likely to arrive with unexpected costs, including deleterious side effects. One expects opposition. The intensity of opposition may fade, however, if the undesirable characteristic is not overly burdensome, if it is accompanied by significant advantages, and if robust numbers of people are profiting from selling whatever product it accompanies.

Expect money to become concentrated and public regulation to constantly play catch-up

Insert here your basic lesson in capitalist agricultural economics. This summary is typical of a widely shared, uncontroversial assessment among experts from the left and the right:

> The U.S. agricultural economy is highly concentrated in the hands of too few processors of major agricultural products. In beef, chicken, pork, seed, and some grains, four or fewer firms so dominate the market that competition is insufficient. Dangerously high levels of buyer market power i.e., monopsony power, prevent America's food producers from receiving an appropriate and

necessary fraction of the retail food dollar. (Domina and Taylor, 2009; cf. GAO-09-746R Concentration in Agriculture, 2009) Wealth in a capitalist economy seems inevitably to concentrate in the hands of fewer and fewer people or corporations. In democratically controlled welfare-state economies, government policies are instituted to try to prevent oligopolies (too few sellers) and oligopsonies (too few buyers). These regulations meet with varying amounts of success. Expect human enhancements to be good for business. Expect government bodies to have uneven and unstable results as they try to regulate the seemingly inevitable concentration of power in internationally competitive markets.

Expect all human enhancement technologies to be tested first on animals

Little needs to be said about this claim. Laws currently require animal testing of drugs, vaccines, other biological and medical devices, to establish safety and efficacy before the product can be tested on humans in clinical settings. If anything, these requirements will become even more stringent as human enhancement technologies begin to come down the pipeline.

Expect every human enhancement technology perfected on an animal to be used on a human

Nor need we say much about this claim. To my knowledge, no drug, vaccine or device that has successfully completed animal trials has not been tested on humans. These include diphtheria toxin, insulin as a treatment for diabetes, halothane as a general anaesthetic, heart valve transplants, and vaccines against polio, leprosy, syphilis and whooping cough (Darling and Dolan, 2007; LaFollette and Shanks, 1996). Many drugs, vaccines and devices shown to work safely in animal tests fail in human clinical trials because they endanger human health or are ineffective. But among those products that have passed clinical trials, it is difficult to find anyone that did not eventually come to market. So, when we have ethical worries about a potential human enhancement, the time to regulate is *before* it is tested on animals. Once the product has passed Phase III trials – the last step in testing on experimental human subjects – the proverbial horse will be out of the barn, as they say.

I know of no technology perfected on so-called animal models and subsequently shown to be safe and efficacious in humans that was not eventually used on humans. I see no reason to expect human enhancement technologies to proceed along a different path.

These six lessons add up to one observation: it is almost certain that experiments will be proposed in which animals are the subjects of surprising interventions. Some of these interventions will seem to many people to be instances of clear abuse. Which experiments should we allow? On what grounds should experiments be judged? Three answers are possible. At either extreme are: 'Permit no experimentation on animals' (animal rights) and 'Permit all experimentation' (speciesism). In the next three sections, I criticise the two extreme views and propose a third.

3 Animal research abolitionists: 'Animals have moral rights and enhancement research is prohibited'

Animal rightists typically oppose all animal experimentation that is not necessary for the welfare of the animal experimented upon. On this view, research aimed at intentional genetic enhancement of animals (IGEA) will be difficult to justify. If animals are, as Regan (1983) argues, subjects-of-a-life with inherent value, then we are not entitled to treat them as means to ends. I will argue, however, that the practical conclusion (no such research) does not follow from the basic principles of animal rights, principles I endorse.

Regan argues that all subjects-of-a-life (henceforth 'subjects') have the right to be treated with respect because they have inherent value (Regan, 1983). Subjects possess a range of mental capacities including sentience, preferences, interests and desires, memory, self-consciousness and a sense of the future; they value their own lives. What happens to subjects matters to them because there is something it feels like to be them, and they exercise a certain amount of control over themselves. Being able to exercise control over one's life means that one's decisions can affect one's welfare, and influence whether one's life goes well or badly. Subjects, therefore, have at least a *prima facie* right that others not interfere with them – try to usurp the subject's control over them – as long as subjects pursue interests that do not harm others.

The particular interests of different subjects make their lives worth living for different reasons. These differences between subjects make lives valuable for different reasons – for reasons determined by the subject itself. The most basic moral right of a subject, therefore, is the right not to be abused in order that others may benefit. Regan holds that many animals – including, at a minimum, all adult mammals – are subjects of a life. Harming these animals for utilitarian reasons – that is, on the basis that others may benefit from the harm done to the animals – is illegitimate because it is a violation of a subject's inherent value. Therefore, the presumptive position of an animal rights ethics is that a proposal to subject animals to painful experiments in order to enhance others is not morally permitted. Those who are subjects of lives cannot be sacrificed or traded-off for the purpose of achieving some allegedly greater good for other subjects of lives. Everyone who has inherent value has it equally. Like humans, animals may not be harmed in order to bring benefits to others.

Regan believes that an abolitionist understanding of animal rights forbids altogether practices such as raising animals for their wool, hunting them for their meat, and using them for entertainment in rodeos and zoos. Writing that 'you don't change unjust institutions by tidying them up', Regan further demands 'the total abolition of the use of animals in science' (Singer and Regan, 1985). Regan's commitment is, as he puts it, to *empty* cages, not *bigger* cages (Regan, 2005). All abuses of the rights of animals in agriculture and research are offensive. It would seem to follow, therefore, that no animal enhancement research should be permitted on traditional animal rights grounds.

I interpret the implications of animal rights differently. On a reformist reading of animal rights, enhancements of animals performed genuinely for the animal's own sake are permitted, and perhaps obligatory under certain highly circumscribed conditions. Reformers agree with abolitionists on the implications of the theory for agriculture: we agree, for example, that people in developed countries generally are not permitted to slaughter animals for meat. And there is overlap in some judgments about the use of animals in research; we agree, for example, that there is a presumption against harming animals to *enhance* humans. But, I will argue, the reformist view allows the use of animals in agriculture and research if and only if the animals are raised in environments where they may pursue their desires, have none of their rights violated and be permitted to live out their lifespans. A further condition would be that we would have to collect whatever by-products we were interested in (e.g., eggs, milk, wool) with a minimum of disruption to the animals' lives. Forms of animal-friendly agriculture are, given current institutional arrangements, economically unfeasible but they are not conceptually impossible. Under different cultural and dietary expectations, and reformed agricultural regimes, we might not only be able to use animal by-products humanely in agriculture. We might also be able to use genetic enhancements to improve animals' lives from *their* point of view.

If you are sceptical about the truth of either of these claims – that we could gather by-products from animals without harming them, or that scientific procedures can be imagined that would genuinely enhance animal lives – you are not alone. Both claims seem counter-intuitive given the definition of animal rights and what we know about the lives of animals. But allow me to set these practical truths aside for the moment and pursue the idea in principle. I want to describe three research projects that, one might think, could be justified on animal rights grounds: smarter mice, less trusting monkeys and more relaxed chickens. I will describe them in more detail in the subsections that follow. But first let me say where I am going with these cases. None of the three examples, I argue, can be justified on abolitionist grounds. However, one can in principle imagine situations in which a variant of each example *would be* justifiable on animal rights grounds.

Smarter mice

A recent study by J. Rekha and associates in India aims at improving the cognitive performance of two-month-old adult rats on maze-solving problems. After isolating and culturing cells from human nasal polyps, Rekha injected the cells into rat embryos, brought the transgenic mice to term, extracted hippocampal cells from the brains of the newborn mice and inserted them into the brains of a dozen adult animals. After recovery from surgery, the enhanced adult animals (EAs) showed memory and learning capacities that outstripped that of ordinary mice by 10–15%. These results are shown in Figure 3.1.

In three of four experimental groups, Rekha lesioned the ventral subicular region of the rats' brains. In one of the four groups, the rats never recovered normal function (VSL). In a second group, the rats received the transplanted cell lines (VSL + H3-GFP). This group contains our 'smarter mice', which quickly came to outperform expectations and significantly exceeded the speed at which unlesioned normal control (NC) rats learned to navigate new mazes. Figure 3.1 shows that the

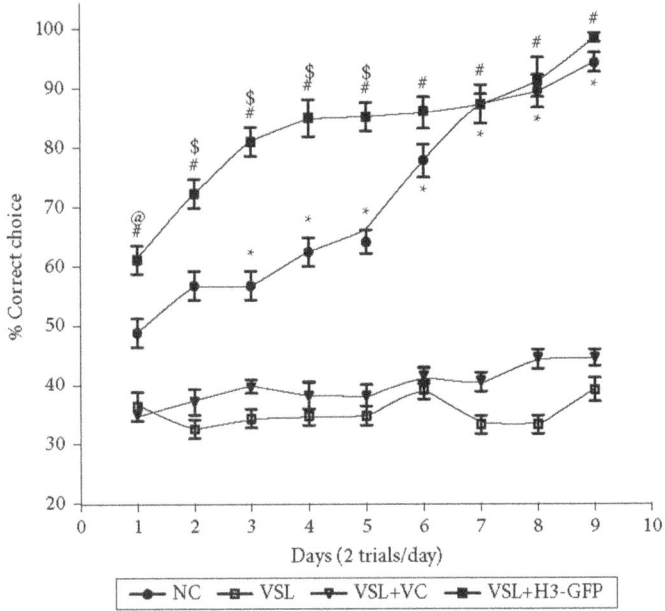

FIGURE 3.1 *(A) Performance of rats in the eight-arm radial maze task*
Source: The graph depicts the performance as percentage of correct choices made by the rats in function of time. Each value represents the mean ± SEM, n = 8 per group. The ventral subicular lesioned (VSL) and VSL + vehicle control (VC) rats showed impairment in task learning and working memory relative to normal control (NC), and VSL rats transplanted with H3-GFP cell lines (VSL + H3-GFP) rats. In addition, the performance of the VSL + H3-GFP group was better than the NC rats from first to fifth day of training and at a comparable level thereafter up to ninth day (Rekha et al., 2009, reprinted with permission of APA).

enhanced mice not only began the maze trial on Day 1 making more correct choices (60%) than normal mice (50%), but after four days were making over 80% correct choices when normals were still at 65%. This 10–15% increase over normal species function represents the degree of enhancement of Rekha's animals.

Rekha enhanced the mice by manipulating somatic cells to produce better working memory and learning. But were the mice improved from *their* perspective? It would appear so, for they were learning to complete tasks – running mazes – and the rapid completion of these tasks was clearly in their interests. Being able to figure out where one is and then to remember the layout of one's environment is clearly vital to a rat's welfare.

Was the enhancement worth the price to Rekha's rats? Well, it might be if the enhancement was achieved without harming the animals or diminishing their skills on other traits. Under these assumptions, the lives of the animals would be better *all things considered*, and it is difficult to object on animal rights grounds. For, as Robert Streiffer (2005) remarks about the prospect of a primate whose mental states have been intentionally enhanced with human brain cells and no other changes, 'How could the animal complain?'

That said, we must face reality. All of Rekha's mice were harmed in the process. They were (just for starters) held in cages for their entire lives, handled by technicians in preparation for surgery and subjected to post-operative recovery. During surgery, chemicals were injected into their brains that caused lesions and impaired cognitive performance. As Rekha's goal was to understand the effects of neurodegeneration afflicting human patients with diseases such as Parkinsonism, he began his experiment by damaging some of the rats' hippocampus formations. Clearly, these rats were worse-off; they took about a third *more* time than the dozen undamaged controls to solve the eight-arm radial maze tests and Morris water maze tasks by which their mental acuity was assessed.

Projects such as this one suggest one might enhance rodent mental performance by stimulating release of growth factors and improving neural generation and brain plasticity. The knowledge gained in the animal experiments might in turn help to produce a treatment for Parkinson's in humans. Eventually, the knowledge might lead to human mental enhancements, too. But let us return to the present again. Rekha's animals were not enhanced. Nor were they used in ways permitted by animal rights principles. Some harms done to them (cages, surgery, recovery) we have already noted. Note, in addition, that whatever mental superiority the enhanced animals had over controls was but a temporary advantage. After five days, unmanipulated normal rats caught up to the enhanced rats in maze-running ability. True, the animals we may presume were anesthetised and given postoperative care as required by law. And yet these procedures themselves involved the immobilisation of the animal's head on a surgical apparatus before its brain was injected with solutions of ibotenic acid (Rekha et al., 2009). From the animal's perspective, it would surely seem that, whatever temporary slight enhancement in problem-solving was achieved by the experiment, it was outweighed by the abuse the animals suffered.

Perhaps we might one day be able to produce smarter mice without harming the animals. Such a research programme might be justifiable by reformist principles. The Rekha experiment, however, is clearly not such a programme. Let us look for another example then. Consider a case in which genetic enhancement technology ensures the survival of a species. Would animal rights reformists allow a research project aimed at saving a species from extinction if the only pains inflicted on animals were minor surgical procedures?

Wilier monkeys

The survival of a species indigenous to Southeastern Brazil is threatened by poachers who are aided by the non-aggressive and non-territorial disposition of the muriqui monkeys they hunt. Suppose we could enhance the monkeys' chances by darting them with an anaesthetic and microinjecting a sex cell altering gene that increases testosterone and leads to greater aggressiveness and sharpened fight-or-flight tendencies. Would we have enhanced the monkeys' lives all things considered? Absent the technology, we suppose, the animals will become extinct. With it, however, they will survive for generations to come. This seems to be a clear case of a morally acceptable use of animal enhancement.

One might argue, further, that this research programme is consistent with animal rights theories. Yes, animal rightists generally have a presumption against intervening in the lives of animals, either in the wilds of Brazil or in scientific research labs. But if an animal species cannot survive without our intervention in the lives of a few individuals, animal rights philosophies might be construed to require choosing species survival.

The reason is that the case presents a conflict of rights; the rights of individual muriqui monkeys not to be darted, anesthetised and microinjected *versus* the rights of future generations of monkeys to exist at all. How would a believer in animal rights resolve such a conflict? The answer may be found in Regan's worse-off principle which applies to cases in which two very different disvalues are at stake. The two values are, on the one hand, the disvalue of being temporarily incapacitated by being shot with a dart gun *versus* the disvalue of not coming into existence at all. The worse-off principle follows from the principle of respect and instructs us what to do in conflict cases (Regan, 1983): *Where non-comparable harms are involved, avoid harming the worse-off individual.*

Think of the muriqui case as a trolley-problem, a forced choice in which we must decide either to take no action – with the consequence that muriquis will become extinct – or to anaesthetise and microinject two muriquis – with the consequence that each animal suffers a 20-minute headache and the species will not become extinct. As non-existence is, on one account, worse than a temporary headache, worse-off requires us to choose the headaches for two animals over the extinction of their species. Assuming that non-existence is non-comparably worse for the muriqui species than a headache is for two animals, worse-off allows us to conduct IGEA in order to produce the less trusting monkeys.

Sounds good in theory, but let us look closer as the practical issues and technical uncertainties are troubling. Primate species have evolved over time to exhibit ranges of behaviours, ranges supported by phenotypic plasticity. Indeed, all mammals seem to have the ability to adjust to a variety of new factors, including changes in kin relations, predator strategies, food scarcities and competitive neighbouring groups. By attempting to fight extinction using the means of genetic engineering, humans might well make things worse, interfering in evolutionary processes that might have hidden resources to handle the challenge. Note the practical difficulties. Genes do not map in any one-to-one relationship on to behaviours, so even if our dart gun scheme to insert a gene is successful, we should have little confidence that we will be able to change the animal's phenotype by deleting a gene or inserting a mutation. Given the unpredictability of the co-evolution of genes and environment and the virtually incomprehensible array of factors we would have to control, it seems highly unlikely that a single genetic change or two could ensure the muriqui's survival. So, even when we carefully try to circumscribe the features of a case so as to depict a situation in which genetic enhancement would immediately serve the good of an animal overall, we still find ourselves, alas, relying on highly questionable assumptions about our ability to control the cascade of events leading from genotype to phenotype. In sum, actual cases of IGEA that are defensible by animal rights principles are imaginable in theory even if we cannot describe any plausible examples at the moment.

Mellower chickens

The welfare of chickens is relative to their environment. In large-scale production facilities, hens are often overcrowded and, as a result, suffer from stress. In 1985, Ali and Cheng noted that farmers could reduce the incidence of cannibalism by selecting for genetic blindness in the birds. Now, if only one bird was blind in the flock, it would have a problem. Incapacitated, it would be an easy target for more aggressive birds. However, if *all* birds were blind, none of them would be able to attack others because they would be unable to see them, much less pick out the weakest members. Ali and Cheng (1985) found that when the entire flock was blind, hens seemed happier. They produced 12 per cent more eggs even though they consumed less feed. They 'were less active, had better feather coverage, and were perhaps under less stress than sighted ones' (p. 789). If blind chickens are under less stress than sighted ones, why not resolve animal welfare problems in the chicken industry by allowing farmers to use only blind chickens? (Rollin, 1995; Comstock, 2000b; Varner, 2012; Sandoe et al., 1999).

Now, according to animal rightists, blinding chickens is obviously not an enhancement of the animal. The very idea has the moral calculus all wrong. For animal rightists, we ought not to address the ills of factory farming by depriving animals of one of their senses. Rather, we ought ideally to cease the practice altogether and, at a minimum, change the environment that is causing stress in the first place.

But imagine that these egg-laying birds are part of my ideal futuristic animal agriculture. All of their needs are met, they are allowed to live out their normal lifespans until they die of natural causes, and their lives are by all accounts happy lives. Happy, that is, except for one problem. They are stressed by seeing other birds in such close proximity. To resolve this source of unease, then (and here is the part where we must use our imaginations), we postulate a poultry scientist proposing a project to produce genetically engineered blind birds. With this enhancement, the hens' lives will be better than any hens in history. So we imagine.

The scenario presents a dilemma for animal rightists. Given the assumptions, no rights are being violated, so the experiment seems permissible. And, on the one hand, we might think that the chickens themselves would be willing to trade in their sight for a stress-free life if doing so really meant – as we are assuming – that they would be better off. Yet, on the other hand, a concern for the animal's integrity, for what some call its 'nature' (and what Bernie Rollin following Aristotle calls its '*telos*'), inclines us to think that such a trade-off would be inhumane, unthinkable. *Blind* a chicken to make it *happier*? Does that phrase even make sense?

Let us summarise our progress so far. There appears to be no current lines of genetic enhancement animal research that are defensible by the lights of abolitionism. However, by extrapolating from current research, we can imagine enhancement research that would be acceptable, for *by definition* the envisioned research would be good for the animal.

In sum, I have shown that even Regan's abolitionist interpretation of animal rights does not necessarily entail prohibitions on all uses of animals in research. When rights conflict, the abolitionist view permits the killing of an animal whenever, for example, no other choices are possible, and the action is necessary to save human life.[2]

On one end of the animal ethics spectrum, then, we have a view that requires vegetarianism but does not require that we rule *out* all animal research. Let us now explore, at the other end of the line, whether speciesism rules *in* all animal research.

4 Speciesists: 'Animals lack rights and all enhancement research is *permitted*'

Speciesists hold that animals lack moral rights. Speciesists are anthropocentrists and argue that whereas inflicting pain on sentient non-human animals may count negatively in the moral calculus, the painless termination of an animal life does not, especially when one animal can be replaced with another. Speciesists hold that the costs to animals of IGEA, if there are any, are so trivial as almost not to matter. As long as humans are benefitted, or humans are eventually enhanced, it does not matter how many animals are harmed. On this view, all scientifically legitimate forms of enhancement research on animals are permissible.

One way to defend speciesism is to argue that typically developing adult humans (henceforth, the typically developing) are persons with moral standing because we have a special moral-status conferring property, such as a soul, language or rationality. Animals, on the other hand, lack the requisite property, whatever it is, and therefore lack moral standing, much less any moral rights.

Some speciesists hold that persons have moral or natural rights because we are members of the human moral community. This is true even if a human's mental capacities are impaired. Normal adult humans care about other normal humans and sympathise with the plight of the congenitally radically cognitively limited (henceforth, the radically limited). These are sympathies we would not want to lose, virtues we would not want to dilute. By entertaining the idea that we should think of misfortunate humans in the same way we think of animals, do we take a step toward becoming people who mistake serious incapacitation as a reason to be cast out of the moral community? 'The issue', writes Carl Cohen (1986), 'is one of kind'.

> Humans are of such a kind that they may be the subjects of experiments only with their voluntary consent. [...] Animals are of such a kind that it is impossible for them, in principle, to give or withhold voluntary consent or make a moral choice. What humans retain when disabled, animals never had. (Cohen, 1986, p. 866)

Here is how the 'kind argument' goes:

1. To have moral rights, one must be *of the kind* that is able to be what humans distinctively can be, or do what humans distinctively can do.
2. Animals are not of the kind that is able to be what humans distinctively can be, or do what humans distinctively can do.

Therefore, animals cannot have moral rights.

This argument is valid but unsound because both 1) and 2) are false. To see why, consider humans with very severe Autistic Spectrum Disorder, or anencephaly, or disabling microcephaly. The radically limited have moral rights, goes the speciesist argument, because they are *of the kind* that is able to be what humans distinctively can be, or do what humans distinctively do. In this respect they are unlike animals. Here is the reason, then, that animals with more sophisticated and complex mental states than impaired humans lack rights while impaired humans have rights. The reason? The impaired humans are of the right kind.

But what is the kind, we may ask? (McMahan, 2002, *cf.* Nobis, 2004). The answer cannot be that the kind is *the features, whatever they are, that typically developing adults and the radically limited share*. To point in some general direction, shrug, and say that all normal humans and so-called marginal humans share a certain *je ne sais quoi* quality – a quality we cannot specify but is clear to everyone nonetheless – is question-begging. We are owed an explanation of the features that hold the group members together. That explanation must point to one of two features: the physical or the psychological features that the typically developing and radically limited have in common. I follow McMahan in holding that there is no third alternative.

The speciesist may appeal, first, to physical characteristics. The radically limited and typically developing both belong to the species *Homo sapiens*, possess human DNA, have a human mother and father, inherit opposable thumbs and so on. The claim is true, and so is the additional claim that rhesus monkeys and other animals lack these characteristics. However, when supplied as a reason for giving moral-status to cognitively challenged humans but not to monkeys, the appeal is hardly convincing. Surely one's physical characteristics are irrelevant to how they ought to be treated. Science fiction examples such as ET illustrate the point. But so do the physical characteristics of unfortunate humans who may not have a human appearance because of disfiguring diseases such as elephantiasis. Physical characteristics will not support the speciesist's case as they have no more bearing on one's moral status than does one's age, gender or skin colour.

If speciesists do not appeal to physical characteristics, they must appeal to psychological characteristics. The radically limited and typically developing, they might say, both have souls, or use tools, or have language, or act on the basis of reasons, or make music and so on. Clearly, one's psychological capacities are relevant to how one should be treated, but a different problem arises here for the speciesist. Whatever psychological capacity or set of capacities the speciesist names, some humans lack it (McMahan, 2005). Cohen's remark to the contrary, not all disabled humans retain what animals never had. Some humans lack not only the ability to use tools, language and morality. They never had the ability and never will. And some animals have some of capacities named, from tool use to music or proto-language. The only psychological candidate left standing is the soul, and it inherits all of the problems of the question-begging *je ne sais quoi* response. When cognitive impairments are severe, and congenital and untreatable, those cognitively limited suffering from them never have what typicals have. And some animals have properties that the limited lack. Psychological kind arguments fail if limited humans fail to have the mental capacities necessary for moral standing, and if some animals succeed in having those capacities.

Speciesists hold that all humans are morally superior to all animals. However, the facts of human neural diversity make it impossible to set the bar to exclude all animals while including all humans. There are no singular morally relevant properties possessed by all humans and no animals, so we lack good reasons to think that we may discount the pains, frustrations and interest in continued life of generations of experimental animals.

If, as I have argued, speciesism is not defensible, then we should not accept its permissive endorsement of all forms of enhancement research.

Summary: the task of assessing these lines of potential research would be simpler if either of two claims were true. First, if abolitionism was true, and one of the consequences of abolitionism was that all enhancement research is ruled out, then our problem would be solved and all animal transgenic research could be prohibited. But even if we assume abolitionism to be true, as we have seen, it does not prohibit enhancement research – and may in fact require it in cases of conflicting rights. Animal rights theories, in brief, do not in fact entail the conclusion that enhancement research is always wrong. On the other hand, were speciesism true, then virtually all IGEA would be allowable. But speciesism arguably is not true so we cannot give transgenic animal scientists *carte blanche*. Speciesist theories, in brief, are unjust. If neither of these popular views is correct, where do we turn for guidance? In my last section I will argue that the utilitarian animal welfare tradition is a good starting place if it is strengthened with a base line of an intuitive-level system of animal rights (Varner, 2008).

5 Animal research reformists: 'Animals have rights but critical thinking may override them in the interest of the overall good – enhancement research is permitted if it meets stringent requirements'

Other arguments exist for denying consideration to animals used in research and for recognising natural rights for them, and these arguments would have to be examined before reaching any firm conclusions about either position. I believe the arguments surveyed earlier, however, are the strongest arguments on offer, and the fact that each position fails leaves us in a difficult place. It seems we must make decisions on a case by case basis, carefully weighing a variety of conflicting, difficult to measure and sometimes incommensurable values. We must consult a variety of ethical principles, assess the likelihood of utility and disutility of various options and respect the rights of all beings while preventing non-comparable harms to those we might make worse-off by our actions.

A third theory may fill the gap. Defenders of two-level utilitarianism recommend that we teach our children to believe in animal rights and to think in those terms ourselves. Given the self-interested and biased nature of human psychology and the unmitigated constraints on the amount of information we are able to acquire before we must make decisions, however, reality requires that we actually make decisions in daily life as if some deontological theory were true. And yet, whenever rights conflict and we have the time and information required to think critically about what is the best action we can undertake, we should turn off the deontological heuristic, make the necessary consequentialist calculations and, according to R. M. Hare (1981), form our decisions as act-utilitarians.

Utilitarians typically do not believe in rights, but Hare's two-level utilitarianism holds that we ought to act in the ordinary case *as if* rights exist. We ought, that is, to instruct our children about people's and animal's rights, because such beliefs are required to counteract the self-interested tendencies of human beings, the vagaries of our knowledge, the ease of self-deception and the impossibility of gathering, much less weighing, all the relevant facts before pressing decisions are on us. That said, I follow Hare and Varner in thinking that when we have the time and resources to engage in critical thinking, we should always make decisions as act-utilitarians, choosing to perform the act that will maximise the overall balance of good consequences over bad consequences (Varner, 2012).

For utilitarians of Hare's sort, then, the notion of a moral or natural right is a convenient shorthand, summarising rules on which we ordinarily ought to act to achieve satisfactory consequences. Two-level utilitarians recommend that we train ourselves to think and act as if we were deontologists in everyday life because this conceit is most likely to maximise overall good, given our inability to overcome bias and lack of information. The default position, then, is to regard moral truisms as well-founded truths and act more-or-less unthinkingly on them. On these assumptions we should, therefore, train ourselves and our children to think of rules such as *Honour thy father and thy mother* and *Do no harm* as reflecting, for example, virtually inviolable human rights to autonomy and protection from being conscripted into experiments in which one's body is used as a means to a scientist's ends (Comstock, 2013).

Two-level utilitarianism requires sensitivity not only to the numbers of animals involved in research but to the differences among species in, for example, sensitivity to pain. The range of psychological potentials of animals matters on this view, as does a rigorous assessment of the harms likely to be inflicted, and the benefits likely to be realised. The distribution of harms and benefits matters, too. Minor harms to a few individuals of an insensate species, for example, might be easily justified if large benefits will accrue both to the individuals used in the experiment and to large numbers of humans whose lives will be saved. If the experimental animal, on the other hand, is a highly intelligent and sentient primate with well-developed emotions and cognitive states, and the research will harm it significantly while having few benefits other than helping older human males to rejuvenate the growth of hair on their bald heads, then the research will be more difficult to justify.

Current animal welfare law suggests that researchers must carefully weigh many factors, an attitude that is in keeping with the spirit of two-level utilitarianism. Let us begin with a quick review of the legislation in the country I know best, the United States. Laws are an important source of moral reflection for us, and they serve to guide us in the ordinary decisions we must make routinely from day to day. Law represents the culmination of what presumably has been a community's democratic process of deliberation. Professional codes distil the results of hours of trained minds focused on ethical questions.

Policies regulating animal research vary between countries. The code regulating animal research in the United States takes as a point of departure the 1966 Lab Animal Welfare Act (AWA) and the subsequent documents that interpreted and built on it.[3] I will refer to this body of laws, rules and guidelines as the AWA tradition. That tradition acknowledges certain limited legal rights for animals, requiring, for example, that each warm-blooded research animal be given a certain standard of care. Exceptions are allowed if care interferes with 'the design, outlines, or guidelines of actual research or experimentation'. Part of the tradition is a 2002 amendment that excludes from the law's protections all birds, mice and rats specifically bred for scientific research.

All research must be aimed at producing knowledge and advancing human understanding of such processes as health, disease and treatment. Animals should not be used in research where the results are already known or can be obtained by other means, including mathematical modelling or *in vitro* experiments or epidemiological studies. Appropriate safeguards must be in place to eliminate unnecessary harms and minimise necessary harms, including the use of qualified medical and scientific personnel and careful monitoring of subjects during experiments. Experiments must be designed to employ methods that reduce bias while using the smallest number of subjects necessary to produce statistically significant results. These rules are all part of current legislation. Excluded from the Act's provisions are birds, rats, mice and farm animals including 'livestock or poultry used or intended for us as food or fibre [...] or for improving animal nutrition, breeding, management, or production efficiency'.

As developed by subsequent laws and policy statements, the AWA tradition stipulates three Rs: consider alternative designs if using a procedure that causes more than momentary pain or distress (*Refine*); minimise the number of animals used (*Reduce*); and use non-animal techniques such as mathematical modelling and *in vitro* studies whenever possible (*Replace*).

Two-level utilitarians are likely to agree with the inclination of those in the AWA tradition to think that difficult cases in which human benefits must be weighed against costs to animal should be decided on a case-by-case basis at a local level by a diverse group of trained people. How would we know when a human or an animal's life has actually been enhanced? Would we use the measures of classical hedonism or desire-satisfaction theory? Or might there be an objective list of values it is good for a mouse, or pig, or cow life to have? And what would be the norm for assessing how much good could be taken from an animal in the name of delivering goods to humans? Would the standard be the typical functioning of an ordinary member of the species? And if so, how would we determine typical functioning? Would the capacities of wild as well as confined animals of a species count?

These are the questions we would have to answer to decide whether any of the four forms of genetic enhancement is ethically justified. Clearly, we cannot make such determinations in advance. We must look at the facts of each case and expect that well-intentioned people will disagree about the correct answers to these dilemmas. The wisdom of the AWA tradition is to recognise that we, lacking democratic consensus about the right answers, should speak with each other locally before deciding when and how to proceed.

I noted that for those who defend a utilitarian understanding of animal rights, researchers using animals are often justified in following existing codes. I now want to revise my claim substantially, because we have the time and resources to reflect on the codes critically and we can easily see that they have two glaring deficiencies. These deficiencies must be corrected before researchers can trust the codes as reliable heuristic guides to action.

6 Two deficiencies in the animal welfare tradition: ethical review and informed consent

The two most striking differences between the policies regulating the use of humans in research and the use of animals is the centrality of ethical review and informed consent in the responsibilities of Institutional Review Boards, and their utter absence from the charge given to members of Institutional Animal Care and Use Committees.

First, consider ethical review. Animal care and use committees are typically not empowered to assess projects on the basis of ethical criteria (Silverman et al., 2006). However, the worse-off principle provides us with a good reason to consider charging animal committees with the responsibility of explicitly assessing the rights of experimental animals as well as the risks and benefits of the project.

Second, consider informed consent. Animal committees are not allowed to require that experimenters try to discover whether animals will consent. The reason is the presumption that animals cannot be informed about what is about to happen to them, nor can they give or withhold consent about what is about to happen to them. It is true that most animals (e.g., multicellular organisms and insects) cannot be informed about procedures. But all adult mammals are arguably able to understand some human intentions, and they can communicate their feelings through a combination of vocal and behavioural cues when they understand that a human intends to harm them. In the case of the use of human moral patients, the law provides that proxies must be named to act as guardians of the interests of the conscripted. Animal researchers could be bound by analogous principles. And they ought to be so bound. We should empower animal committees to require proxies to represent the interests of sentient animals that researchers are proposing to use in potentially painful research.

For those who think of rights in the way recommended by Hare's two-level utilitarian theory, rights function as a shortcut to good decisions. Researchers should train themselves and others to think of experimental animals as having rights, and generally govern their behaviour in light of this intuitive standard. The reason is that the best set of consequences are those we would be able sincerely to embrace if we had to live out the lives of each one – human and non-human – affected by our action. Those adopting this view will think of experimental animals as having a right to a certain standard of health and well-being.[4]

7 Conclusion

I have argued that we should base policies about using animals to enhance humans on a hybrid utilitarian theory. Generally speaking, we should in our everyday lives approach the issue by thinking that the AWA tradition provides viable guidelines. However, when we think critically, we should form policies in a way that is sensitive to the different interests of different species and to the quantity and quality of good that each individual brings into the world. We must weigh all considerations, permitting uses of animals only if it is clear that doing so will maximise good.

The AWA tradition provides a theory that is sensitive to the rights concerns; it provides a good starting point for the development of better rules. That tradition, however, wants development in two directions: enabling animal use committees to review the ethical dimensions of proposals, and requiring informed consent from animals capable of giving it, should there be any.

If I am right, the provisions of current U.S. law referred to earlier as the AWA tradition are a defensible but defective set of baseline rules. The three Rs are a set of critical threshold policies that might govern using animal models for human enhancement purposes. Thus, for example, all human enhancement research on animals should produce scientific knowledge that aids human understanding and health, and cannot be obtained by other means such as mathematical or *in vitro* models or epidemiological studies. But two-level utilitarianism requires that further safeguards be put in place to eliminate unnecessary and minimise necessary harms to animals, such as the use of qualified medical and scientific personnel and careful monitoring of subjects during experiments. Experiments must be designed to employ methods that reduce bias while using the smallest number of subjects necessary to produce statistically significant results. These rules are all part of current legislation. I have argued, however, that these rules should be supplemented with these additional rules:

1. Decisions about enhancement research protocols should aim to maximise the ratio of the well-being of individual humans and animals with moral standing while minimising harms to them.
2. To respect rule 1 in the face of our tendency for self-interested bias and inability to acquire full information within the time constraints of decision-making, we should train ourselves to act as if we are deontologists, respecting the 'rights' of animals as well as humans

until we are convinced that violating some right is necessary to respect rule 1.
3 Proposals that require violating the rights of experimental animals must present arguments of increasing weight as the moral standing of the animal approaches the moral standing of persons.
   a. The weightiest arguments for violating animal rights will be those that lead to the imposition of the least harms on animals in the service of respecting the rights of the largest numbers of humans.
   b. The least weighty arguments for violating animal rights will be those that lead to the imposition of the greatest harms on animals in the service of achieving enhancements of least importance to the well-being of the fewest numbers of humans.

As we have seen, moving from proposals to use animals to *heal* humans to using them to *enhance* us places a weightier burden of proof on those who want to harm animals. The more trivial the human interest served by enhancement the stronger the presumption that the research is unethical. This does not mean the presumption cannot be overcome, however. Suppose, for example, that we could give future generations of humans a genetic disposition to think of other mammals as having rights (the proposed change would be an enhancement, not a treatment, because one does *not* have a disease if one thinks of animals as lacking rights; to the contrary, that mistaken thought is the result of an error in reasoning, not the result of an illness). If we could reduce the incidence in future generations of people who think that animals lack rights, we might also dramatically improve the lives of billions of future animals. Would researchers be justified in harming experimental animals *now* in order to improve future humans' reasoning capacities if by so doing they could ensure that future animals would be treated more humanely? The answer will hinge on many currently unknown details, of course. I raise the possibility only to underline the fact that utilitarian-inspired defenders of animal rights must oppose many uses of animals in genetic enhancement research – but not necessarily all.[5]

Notes

1 The point at which treatment turns into enhancement is a slippery one (Daniels, 2000, cited by Savulescu, 2006). However, for our purposes, we

need only the roughest idea of the difference between treating disease and enhancing a normally functioning trait. If a patient has anemia resulting from chemotherapy, and one hypothesises that recombinant human erythropoietin (EPO) may relieve the illness, then EPO research done on animals to test this hypothesis will be done in the name of improving therapy. On the other hand, if a healthy person desires to shave a few seconds off of their marathon time and hypothesises that recombinant human EPO will help them to do so, then EPO research done on animals to test this hypothesis will be done in the name of enhancement. The first set of issues, harming animals to heal humans, is much discussed in the literature. I focus here on the latter set of issues, that is, harming animals to *enhance* humans.

2 Admittedly, this consequence does not sound like one Tom Regan would endorse. Regan (1983), known for his abolitionist stance toward the use of animals in science, explicitly denies that his theory leads to the conclusion here described. For further discussion, see Comstock (2000b).

3 By the AWA tradition of interpretation, I have in mind the following sorts of documents: the Animal Welfare Act, 1966, as amended in 1970, 1976, 1985 and 1990; Animal Welfare Regulations; Health Research Extension Act, 1985; Unites States Government Principles for the Utilization and Care of Vertebrate Animals Used in Testing, Research, and Training, 1985; PHS Policy on Humane Care and Use of Laboratory Animals, 1986; 2000 Report of the AVMA Panel on Euthanasia; Guide for the Care and Use of Laboratory Animals; and the NIH Grants Policy Statement.

4 How do we determine that standard? Does justice require that we set the standard by identifying the normal functioning of a member of that animal's species? Or, instead, by identifying each animal's individual capacities? These questions I leave for discussion on another occasion.

5 I thank Gustaf Arrhenius for his commentary on an earlier draft, Simone Bateman and Jean Gayon for editorial suggestions and members of the Paris workshops on human enhancement for helpful discussion.

References

Ali A. and Cheng K.M. (1985) 'Early egg production in genetically blind (rc/rc) chickens in comparison with sighted (Rc+/rc) controls', *Poultry Science*, 64(5) May, 789–94.

Aqua Bounty Technologies Inc. (2010) 'Environmental assessment for AquAdvantage® Salmon, an Atlantic salmon (Salmo Salar L.) bearing a single copy of the stably integrated A-form of the opAFP-GHc2 gene construct at the A-locus in the EO-1α Line', www.fda.gov/downloads/AdvisoryCommittees/.../UCM224760.pdf.

Bostrom N. (2003) 'Human genetic enhancements: a transhumanist perspective', *The Journal of Value Inquiry*, 37(4), 493–506.

Cohen C. (1986) 'The case for the use of animals in biomedical research', *New England Journal of Medicine*, 315, 865–70.

Comstock G. (2000a) 'An alternative ethic for animals', in Hodges J. and Han I.K. (eds), *Livestock, Ethics and Quality of Life* (Wallingford UK and New York: CABI Pub).

Comstock G. (2000b) *Vexing Nature? On the Ethical Case against Agricultural Biotechnology* (Dordrecht: Kluwer Academic Publishers).

Comstock G. (2013) *Research Ethics: A Philosophical Guide to the Responsible Conduct of Research* (Cambridge: Cambridge University Press).

Daniels N. (2000) 'Normal functioning and the treatment-enhancement distinction', *Cambridge Quarterly of Healthcare Ethics*, 9(03), 309–22.

Darling F. and Dolan K. (2007) 'Animals in cancer research', *Lab Animal Europe*, October, http://www.animalresearch.info/en/medical/articles/animalsincancerresearch.

Directorate-General for Communication (2010) 'Special Eurobarometer 341-73.1: Biotechnology' (Bruxelles: European Commission), http://ec.europa.eu/public_opinion/archives/ebs/ebs_341_en.pdf.

Domina D. and Robert Taylor C. (2009) 'The debilitating effects of concentration in markets affecting agriculture' (Lincoln NE: Organization for Competitive Markets), http://www.competitivemarkets.com/debilitating-effects-of-concentration-report/.

GAO-09-746R Concentration in Agriculture (2009) 'Agricultural concentration and agricultural commodity and retail food prices: briefing for congressional staff' (Washington: U.S. Government Accountability Office), http://www.gao.gov.

Government of Canada, Health Canada (2002) 'Report of the Canadian Veterinary Medical Association Expert Panel on rbST [Health Canada, 1998]', http://www.hc-sc.gc.ca/dhp-mps/vet/issues-enjeux/rbst-stbr/rep_cvma-rap_acdv_tc-tm-eng.php.

Hare R.M. (1981) *Moral Thinking: Its Levels, Methods and Point* (New York: Oxford University Press).

LaFollette H. and Shanks N. (1996) *Brute Science: Dilemmas of Animal Experimentation* (London and New York: Routledge).

Lee S., Barton E.R., Lee Sweeney H. and Farrar R.P. (2004) 'Viral expression of insulin-like growth factor-I enhances muscle hypertrophy in resistance-trained rats', *Journal of Applied Physiology* (Bethesda MD), 96(3), 1097–104, doi:10.1152/japplphysiol.00479.2003.

McMahan J. (2002) *The Ethics of Killing: Problems at the Margins of Life* (Oxford: Oxford University Press).

McMahan J. (2005) 'Our fellow creatures', *The Journal of Ethics*, 9(3/4) January 1, 353–80, doi:10.2307/25115832.

McPherron A.C. and Lee S.-J. (1997) 'Double muscling in cattle due to mutations in the myostatin gene', *Proceedings of the National Academy of Sciences*, 94(23) November 11, 12457–61.

Nobis N. (2004) 'Carl Cohen's "kind" arguments for animal rights and against human rights', *Journal of Applied Philosophy*, 21(1), 43–59, doi:10.1111/j.0264-3758.2004.00262.x.

Overall K.L. and Dunham A.E. (2002) 'Clinical features and outcome in dogs and cats with obsessive-compulsive disorder: 126 cases (1989–2000)', *Journal of the American Veterinary Medical Association*, 221(10) November, 1445–52, doi:10.2460/javma.2002.221.1445.

Overall K.L., Hamilton S. and Chang M. (2006) 'Understanding the genetic basis of canine anxiety: phenotyping dogs for behavioral, neurochemical, and genetic assessment', *Journal of Veterinary Behavior: Clinical Applications and Research*, 1(3) November, 124-41, doi:10.1016/j.jveb.2006.09.004.

Pursel V., Pinkert C., Miller K., Bolt D., Campbell R., Palmiter R., Brinster R. and Hammer R. (1989) 'Genetic engineering of livestock', *Science*, 244(4910) June 16, 1281–8, doi:10.1126/science.2499927.

Regan T. (1983) *The Case for Animal Rights* (Berkeley: University of California Press).

Regan T. (2005) *Empty Cages: Facing the Challenge of Animal Rights* (Lanham MD: Rowman & Littlefield).

Rekha J., Chakravarthy S., Veena L.R., Kalai V.P., Choudhury R., Halahalli H.N., Alladi P.A. et al. (2009) 'Transplantation of hippocampal cell lines restore spatial learning in rats with ventral subicular lesions', *Behavioral Neuroscience*, 123(6), 1197–217, doi:10.1037/a0017655.

Rollin B.E. (1995) *The Frankenstein Syndrome: Ethical and Social Issues in the Genetic Engineering of Animals* (Cambridge: Cambridge University Press).

Salgirli Y. and Dodurka T. (2011) 'The first report of self-directed aggression in a stray dog in Turkey', *Journal of Veterinary Behavior:*

Clinical Applications and Research, 6(6) November 1, 351–4, doi:10.1016/j.jveb.2011.03.004.
Sandoe P., Nielsen B.L., Christensen L.G. and Sorensen P. (1999) 'Staying good while playing god – the ethics of breeding farm animals', *Animal Welfare* (South Mimms UK), 8(4), 313–28.
Savulescu J. (2006) 'Justice, fairness, and enhancement', *Annals of the New York Academy of Sciences*, 1093(1), December 1, 321–38, doi:10.1196/annals.1382.021.
Silverman J., Suckow M.A. and Murthy S. (2006) *The IACUC Handbook* (Boca Raton FL: CRC Press).
Singer P. (reply by) and Regan T. (1985) 'The dog in the lifeboat: an exchange', *The New York Review of Books*, April 25, http://www.nybooks.com/articles/archives/1985/apr/25/the-dog-in-the-lifeboat-an-exchange/.
Streiffer R. (2005) 'At the edge of humanity: human stem cells, chimeras, and moral status', *Kennedy Institute of Ethics Journal*, 15(4), 347–70.
Streiffer R. and Gavrell Ortiz S. (2010) 'Animals in research: Enviropigs', in Comstock G. (ed.), *Life Science Ethics*, 2nd edn (Dordrecht: Springer), pp. 405–22.
Varner G.E. (2008) 'Utilitarianism and the evolution of ecological ethics', *Science and Engineering Ethics*, 14(4) October, 551–73, doi:10.1007/s11948-008-9102-5.
Varner G.E. (2012) *Personhood, Ethics, and Animal Cognition: Situating Animals in Hare's Two-Level Utilitarianism* (Oxford: Oxford University Press).
Zhou Z., Sheng X., Zhang Z., Zhao K., Zhu L., Guo G., Friedenberg S.G. et al. (2010) 'Differential genetic regulation of canine hip dysplasia and osteoarthritis', *PLoS ONE*, 5(10) October 11, e13219. doi:10.1371/journal.pone.0013219.

4
Sex Hormones for Animals and Humans? Enhancement and the Public Expertise of Drugs in Post-war United States and France

Jean-Paul Gaudillière

Abstract: *This chapter focuses on the questions raised by the administration of DES to animals for food production, within the context of the DES medical affair and the public debates about the regulation of sex hormone prescription. The contrast in the development of this practice in the United States and France originates in the different roles public expertise played in each country. The chapter concludes that the different meanings thus given to human and animal enhancement – improvement versus augmentation – rest on a common ground. Agriculture and medicine are not two worlds apart: the same tools, techniques and sometimes personnel circulate between the two, thus making it possible to enhance both type of bodies at the boundary between the normal and the pathological.*

Bateman, Simone, Jean Gayon, Sylvie Allouche, Jérôme Goffette and Michela Marzano, eds. *Inquiring into Animal Enhancement: Model or Countermodel of Human Enhancement?* Basingstoke: Palgrave Macmillan, 2015.
DOI: 10.1057/9781137542472.0009.

In April 1973, the Food and Drug Administration (FDA) took the radical decision of banning diethylstilbestrol (DES) implants in cattle and other livestock, after having prohibited the use of DES in animal feeds one year earlier.[1] Cancer in humans was the official motive for this unusually rash action, which made an additive that had been used in animal feeds for more than 20 years as a growth enhancer disappear from the husbandry scene. The decision was immediately contested in court and even if a federal judge finally approved the FDA ban in 1978, illegal uses of DES in agriculture remained an issue until the mid-1980s.

The story of DES in the United States has been explored from many different perspectives. Most studies have addressed the uses of the drug in medicine, focusing on the controversy, which surfaced in the 1970s when it was discovered that the massive prescription of DES to pregnant women in the 1950s–1960s to prevent spontaneous abortion resulted in cancer and anomalous development of the reproductive track in the daughters of treated women. One group of studies has looked at the unfolding of this medical drama, focusing on the lessons to be drawn from the affair, either from the perspective of medical practice or of its regulation (Apfel and Fisher, 1984; Meyers, 1986; Dutton, 1988; Pfeffer, 1992). Another group, mostly historians and sociologists of medicine, has investigated the medical uses of DES to include them in the long history of sex hormones and gynaecology, while more recent studies on gender have analysed the role played in the crisis by the then-emerging Women's Health Movement (Bell, 1980; Marks, 2001; Morgen, 2002). A much smaller group of studies dealing with the 'meat and DES' controversy has emphasised its two distinct contexts: the rise of industrial agriculture; and the development of the environmental and consumer movements in the 1960s and 1970s (Shell, 1984; Marcus, 1986; Rifkin, 1992). This literature on agricultural DES shares a strong interest in the problem of 'regulatory capture', namely, the different ways in which political and economic interests have made possible both its extended use and – when social movements grew in importance – its final ban.

This chapter, which stems from the last of these scholarly trends, does not take agricultural DES as a case study for analysing the impact of consumer politics on the relations between industrial agriculture, public health and the surveillance of dangerous substances. Its aim is to take the DES crisis as a point of entry into the question of animal and human enhancement. The massive prescription of DES to pregnant women was an off-label use of this molecule and could, as such, be viewed as an (unfortunate) attempt at enhancement, taking pregnancy as a problematic condition that could be improved. Indeed, the risk of miscarriage was not initially among the pathological indications retained by the FDA to authorise the marketing of DES; it was an off-label extension of prescription patterns. Although the research backing such use had never been sufficient to build consensus in the 1950s and 1960s, the practice was nonetheless deemed safe because DES was an analog of natural oestrogens. In practice, its use became so widespread that many women without any specific history of miscarriage were being prescribed DES as preventive medication. In parallel, the uses of DES in agriculture together with antibiotics in order to accelerate the growth of farm animals, from chicken to cattle, may be viewed as a paradoxical case of enhancement: it originated, not in attempts to ameliorate animal life, but in a technological race to improve human health by increasing the consumption of meat through its cheap mass-production. This enhancement of human health was however indirect, since it implied interventions aimed at the augmentation of the animal body – that is, changes in its weight, muscular constitution and appearance – so as to boost an animal's meat production capacity, increase meat consumption and thus (supposedly) improve human life. These two examples of enhancement seem utterly different and are rooted in distinct social worlds: one case targets the medicalised reproduction of humans; the other, the industrialised production of enlarged animals for the sole benefit of humans. Moreover, these two examples highlight different ways of understanding enhancement: as an improvement and as an augmentation. However, viewed from a practical point of view, agriculture and medicine are not two worlds apart: the same tools, techniques and sometimes personnel circulate between the two, thus making it possible to 'enhance' both types of bodies. This chapter will therefore focus on the use of sex hormones as a practice common to animal and human enhancement and on the ways in which the DES problem was constructed in public spaces with respect to competing views and representations of enhancement, both human and animal.

The argument will be presented in three steps. The first section will recall how DES first entered agricultural practice during the 1950s in the United States, and how its uses rapidly became controversial. The main focus will however be on the form of the 1970s crisis as a specific and decisive feature of the US context, namely, the conjunction of this agricultural controversy and the medical scandal originating in the cancers and malformations diagnosed in women who were later collectively identified as the 'DES-daughters'. The second section will explore the form of public expertise that characterised the US debates: given their roots in consumer politics, debates focused on the ways in which the risks were being evaluated, making it possible for congressional hearings to approach the adverse consequences of DES use both in medicine and in agriculture. The last section presents the French configuration of these debates, which provides an alternative view to the US focus on sex hormones as a common feature of both animal and human enhancement: discussions of the medical and agricultural uses of DES in France remained unrelated, because animals were simply left out of all debates on failed attempts to enhance human reproduction.

1 Agricultural DES in the United States: a contested growth enhancer

Charles Dodds first synthesised DES in 1938.[2] Working for the Medical Research Council, the British chemist did not patent the process, even though the molecule revealed promising properties. DES did not present a structural similarity with natural sex steroids, but proved to be a potent analog of oestrogens, mimicking most, if not all, of the latter's effects in the animal assays then in use for assessing the potency of female hormones. Cheap and easy to produce, DES rapidly became a substitute – and competitor – for industrially made oestrogens, namely, those purified by pharmaceutical firms from the urine of pregnant mares or synthesised from cholesterol. Initially, DES was used in gynaecology as a therapeutic agent in a variety of indications usually handled with oestrogens: infertility, sexual-cycle disorders, uncontrolled bleeding, absence of menses or problematic menopausal symptoms.

The FDA authorised DES for the US medical market in 1941. As reported by S. Bell, the 'synthetic estrogen' played an exemplary role in the history of the drug agency as it was one of the first compounds to be approved according to the procedures defined in the 1938 Food, Drug, and Cosmetic Act. The act partially transferred burden of proof to the industrialist. A manufacturer seeking authorization was mandated to document the safety of its product, but the new law did not define any type of acceptable evidence or test. Approval was granted for the basic gynaecological indications, that is, amenorrhea, menopausal symptoms and infertility. Off-label uses of what was perceived as a potent oestrogen rather than an artificial analog boomed in the 1950s, the most important of these – in terms of prescription numbers – being the management of the risk of spontaneous abortion during pregnancy. Indeed, it had become widely accepted that DES could act as a 'replacement' therapy for a risk attributed to oestrogen deficiency (Bell, 1980). If large uses of DES were never considered trivial, only a small minority of physicians remained concerned about the fact that, as early as the 1930s, laboratory experiments with mice had shown that oestrogens in general, and DES in particular, could induce tumours in healthy animals.

Parallel to this widespread medical prescription of DES in the 1950s, and following pioneering work by animal nutritionists at Iowa State College under W. Burroughs' lead, the use of DES was extended to agriculture. Burroughs and his colleagues discovered that DES, given in minute amounts, accelerated the growth of calves (Marcus, 1986, ch. 1 and 2).³ The idea of adding DES to industrially prepared premixes was patented by the university and licensed exclusively to the pharmaceutical firm Lilly. The process was a huge success. Lilly sublicensed it to a few dozen companies. As a result, within two years, more than six million calves were being fed nutrition containing DES.

FDA approval for the technique was obtained easily. Rather than providing evidence for the non-toxicity of the additive (not required for animal drugs), the Iowa nutritionists argued that no residues of DES could be found in the carcasses of animals fed with the growth-enhancing premixes. The risk for the consumer was therefore nil. However, to be on the safe side, the FDA imposed a withdrawal period of two days between the last treatment and the slaughter of the animal.

Physicians interested in environmental carcinogens – an issue gaining visibility in the late 1950s – were the first to link agricultural DES and the adverse properties of oestrogens. In 1956, J. Smith – a New York state physician – argued at the annual meeting of the Association of Official Analytical Chemists that DES should be considered as a 'cancer threat' given the massive experimental evidence of its carcinogenic effects. Barely taken into account during the meeting, the argument nonetheless made its way to the *New York Times* (29 January 1956) and to the US Congress.

The historian Alan Marcus rightly emphasised that the invention of DES-supplemented feeds was a typical event of the post-war transformation of US agriculture, which had generalised the industrial farming model initially experimented with poultry. Modern agriculture meant not only scaling up herds, and intensive rather than extensive rearing, it was also synonymous with forms of 'scientific farming' that relied on a controlled input/output balance. These changes implied the use of standardised feed and nutrition supplements in order to increase the productivity of bodies perceived as meat-production machines or to boost the health of animals kept in dense populations (often indoors), making them highly susceptible to the spread of contagious diseases. Oestrogens thus entered the meat chain along with the massive use of vitamins and antibiotics.

This transformation of agricultural practices became a matter of concern and an issue of public debate in the 1960s. The trajectory of DES was rapidly affected by the emergence of critical voices associated with the consumer movement. Although their origins can be traced back to the 1930s, the heterogeneous organisations defending 'consumer rights' experienced rapid growth and radicalisation during the 1960s, a phenomenon often attributed to the favourable climate generated by the economic expansion of the post-war decades (Silber, 1983).

The enactment of what came to be known as the Delaney clause in 1958 was a critical event in the polarization of debates on the quality of food (Marcus, 1986). This amendment to the FDA Act, introduced by Congressman Delaney, stated that no food additive found to induce cancer either in animals or in humans could be authorised (*US Statutes*, 1958, pp. 1784–9; *New York Times*, 1958). The measure was adopted despite the FDA's opposition. The FDA considered this broad ban as both prejudicial to the practice of productive agriculture and impossible to enforce. The approval of the clause was not only a symptom of the mounting influence of consumer activism, but – as testified by the parallel drawn during the Delaney hearings between DES and dichlorodiphenyltrichloroethane, the widely used insecticide commonly known as DDT – was also one of the first legislative consequences of an environmental movement that was beginning to benefit from a powerful middle-class constituency and that no longer restricted its actions to conservation issues but also embraced concerns about the dissemination of chemicals in the environment.[4]

For ten years after it was enacted, the Delaney clause was constantly under fire during the debates in Congress on industrialised food. Farmers, industrial-feed producers and drug companies joined forces in attempts to have the clause repealed, arguing that it would ban all sorts of chemicals that could be used for the betterment, preservation and conservation of food. They claimed that animal tests could not be trusted, upholding that it would always be possible to find models and circumstances under which any chemical could become a carcinogen; they therefore explored ways of repealing the clause. In 1962, agricultural lobbying succeeded in inserting a short addition to the amended Food and Drug Act, which stated that a substance could be authorised, even if proven a carcinogen in the laboratory, provided it did not harm the animals and left no residues in the meat for human consumption.[5]

Controversies over the dangers of DES grew in the context of the 1960s widespread critique of government, industrial capitalism and established authorities. The public life of DES was to a large extent determined by the power struggle opposing the alliance for industrial agriculture and the loose front connecting the social movements of the 1960s and their Democratic allies. Best-selling books such as William Longgood's *The Poisons in Your Food* (1960), Rachel Carson's *Silent Spring* (1962) and James S. Turner's *The Chemical Feast* (1970) accompanied the rise of environmental and consumer organisations. These new waves of activism weighed heavily in two different areas. First, Democratic as well as Republican governments conceded that stronger regulation was necessary to tame the market. In the early 1960s, President Kennedy mandated the official recognition of four consumer rights: the right to safety, the right to be informed, the right to choose and the right to be heard. In parallel, he appointed a Citizen's Advisory Committee whose role would be to conduct hearings on the aims and means of the FDA, and provide recommendations for its reform. Some of these were incorporated into the 1962 Act, strongly reinforcing the agency's role in evaluating both the toxicity and the efficacy of therapeutic agents in human medicine. Second, some of the reports and studies supporting the positions of social movements were given nationwide attention. This was the case, for instance, of the reports distributed by Ralph Nader's Centre for Study of Responsive Law, which began to investigate FDA activities in 1968 after its first major success in indicting the car industry (Nader, 1965).

What finally emerged during this first – low key – discussion of the uses of DES was concern over these practices of animal enhancement, understood as augmentation of animal growth, but with an uncertain impact on human health, since DES residues with possible carcinogenic effects could be passed along the food chain.

Most historical work – Marcus's included – on the relations between the DES affair and this particular context has been elaborated in terms of political interests and distrust of scientific expertise. The argument is that reassuring discourses about the limited nature of the cancer risk involved in the use of food additives did not convince the public. Both sides (in favour and against the use of DES) waged a polemical battle, defending their views and interests in the name of science. One side upheld that the social movements were unable to accept the existing facts regarding the limited dangers of DES in meat, while the other side argued that big agro-industrial interests produced ignorance and bad science to mislead the public. In this perspective, experts made statements that went far beyond what was known at the time, building on their social and political commitments.

It would however be misleading to limit our reading of the DES affair to the making of such alliances. Choices regarding values and attitudes toward the medicalization of human life on the one hand and toward the rise of industrial agriculture and the transformation of animal bodies into food-production machines on the other hand were being expressed and were playing a critical role. The convergence of the two DES affairs also raised concerns about the adverse effects of enhancement in humans and in animals, thus bringing up the issue of safety and leaving out a putative alternative framing that would have included the respect of natural bodies and their normal physiological processes or the quality of animal life.

The decisive decade was in this respect not the 1960s but the 1970s, when the links between DES as a drug and DES as food were made, bringing in another layer of conflicting and situated expert public discourses. Medical uses of DES were basically unproblematic until 1971, even though feminist authors would occasionally draw an analogy between the contraceptive pill and targeted gynaecological hormones for their purported carcinogenic properties (Seaman, 1969). The professional 'warning' on DES began locally in 1970, when gynaecologists at the Massachusetts General Hospital reported a surprising series of vaginal tumours.[6] Considered as very rare, this type of cancer had been diagnosed in rapid succession in very young women, a phenomenon which was even more unusual. Following these early observations, Arthur Herbst and his colleagues reinforced their argument that this was a serious public-health issue by focusing on the epidemiological evidence. They selected a retrospective control group by looking at the records of women admitted the same day as their patients, matching age and social groups. The statistical comparison of their records revealed one single significant difference: the girls suffering from vaginal cancer were born from women who had been treated with DES during their pregnancy. The correlation was published in April 1971 in the *New England Journal of Medicine*, after Herbst (Herbst et al., 1971) had sent his data to the FDA. The article alerted the New York Public Health Service, and its cancer-control bureau started to look for similar cases. Having found another five, also correlated with 'DES mothers', they issued a general warning to the state's physicians (Greenwald et al., 1971). It did not take long for the initiative to find its way to the mainstream media. In less than six months, the professional alert had become a national scandal, relayed by public-health authorities and local gynaecologists and widely discussed in the news. It became a topic for congressional hearings and an ongoing motive for concern within the FDA.

2 Food AND drug: congressional hearings and the public evaluation of enhancement practices

The public debates on DES that took place during the 1970s were not restricted to the classical discussion of toxicological facts, such as the production and interpretation of dose-response curves to assess risk. The response provided to the question of what to do with agricultural DES, that is, ban it or keep using it under controlled conditions, depended on how each protagonist considered a series of issues: the relationship – causation or association – between DES and the young patients affected with vaginal cancer, the ability to measure DES residues in carcasses, the meaning of cancer in mice exposed to low doses of DES and the benefits of increased productivity and accelerated animal growth for the 'American consumer' and the economy at large. The sheer number of congressional investigations on one or both uses of DES (ten between 1971 and 1981) testifies to these intertwined concerns regarding the legitimacy and consequences of large-scale alteration of hormonal regulation in humans and animals.

Congressional hearings are worth some interest because of the special role they play in the U.S. regulatory system. Hearings are parliamentary events, in which the work of experts is put on stage in a political context that targets stakeholders. Expert testimonies are gathered to balance and draw the line between science and politics, between what is reasonably well-known and what is not, between sound science and speculation, between the respective merits of various forms of evidence – all of which are considered from the perspective of the interests such knowledge may serve or disserve. A specific feature of U.S. congressional hearings is therefore not the fact that experts are questioned by members of a bipartisan committee with significant power of investigation, but the agreed-upon concept that all 'interested' parties must be brought in to present their views: each party has its own expertise, and may therefore legitimately exert some influence in order to reach a reasonable compromise that balances situated views and power. Political analysts have pointed out that this model of participatory stakeholder democracy emerged during the progressive era as a consequence of the weakness of both Congress and the federal administration, the latter having had to negotiate its action with a distrusting 'civil society' beyond the borders of official 'representation'.

The hearings provided a theatre for the unfolding of new proofs. Thus studies such as those conducted in the 1950s on the efficacy of DES experienced new life (for instance, Dieckmann et al., 1953). It was rediscovered that pregnancy was not among the indications the FDA had initially approved: administrating DES during pregnancy was an off-label practice, resulting from a collective consensus among women's physicians. The political construction of the affair reinforced the view that the FDA had failed to notice and control market-triggered wrongdoings, as its most recent assessment of DES had demonstrated. The National Academy of Sciences (NAS), as part of a general review ordered by the Reform Act of 1962, had re-examined the evidence for efficacy published in the medical literature or presented in the companies' initial applications, and had concluded in 1969 that the prevention of spontaneous abortion was only 'possibly effective'.[7] FDA critics contended that in light of this information, the FDA should have put pressure on the firms to provide new data and, had the firms been unable to comply, should have warned physicians that pregnancy was actually an 'unproven' indication.

The hearings greatly contributed to tightening the link between DES in cattle, residues in meat and the risk of cancer in humans. An example is the first hearing following the publication of the Herbst article on vaginal cancer, which took place in the House on 11 November 1971. Although it had been set up to address the medical alert, the investigating committee headed by Representative L. Fountain, a Democrat, repeatedly raised the question of what the newly discovered human cancers might imply for the agricultural uses of DES. Dr. Herbst himself was directly questioned for his opinion regarding the risks of eating meat from cattle fed with DES.

In his first testimony, FDA Commissioner Edwards tried to play down both the medical alert and its translation into the agricultural idiom (US Congress, House of Representatives, 1972, pp. 49–54 and 76–84). Asked to explain why his agency had not reacted when Herbst first sent his clinical reports in March 1971, before the publication of his article, and why the FDA had left the issue pending until the eve of the hearings, the head of the FDA argued that a correlation was not a demonstration of causality. Herbst had shown an association, but this association needed to be confirmed in order to rule out the intervention of 'confounding factors'. More science was needed before a generalisation of the risk could be acknowledged, especially if a bridge was to be established between the effects of a high therapeutic dosage in humans and the extremely low concentrations of food additives used in animal feeds. Central to Edward's scepticism was the huge difference in the doses administered in each context.

Given the centrality of the dose-effect relationship in the pharmacological and toxicological perspective shared by both the FDA and specialists in animal nutrition, the question of dosage was systematically addressed by each expert testimony, and directly, through computations of inoculated or ingested quantities. It was also dealt with in a less direct way, when discussing the relationship between DES and natural oestrogens. It is not altogether surprising that feeders and animal nutritionists argued that DES was an oestrogen like any other, thus explaining that its carcinogenic potency was just a manifestation of its ability to mimic the properties of sex hormones. Consequently, they considered that DES quantities should not be considered in isolation, but should be discussed as a component of a global oestrogenic pool that included much higher quantities of normal oestrogens, in the bodies of both supplemented cows and women. This position was echoed by the representative of the Middle West participating in the House committee, but also by prominent endocrinologists. For instance, Mortimer Lipsett, scientific director at the National Institute of Child Health and Human Development, confirmed this similarity between DES and natural oestrogens. As the situation evolved, however, the equivalence became looser. During the last hearings, held in 1978, the industry adopted the opposite position:

they separated DES from the other oestrogens. Given that the former was generally considered dangerous and was on the verge of being definitively banned, it then became good policy to preserve other oestrogens from an equivalent threat so that they could replace DES as growth enhancers.

In reading through the series of both written and oral expert testimonies presented at House and Senate hearings during the 1970s, three types of discourse can be identified that link the different levels of the DES controversy. The first is an epidemiological discourse focusing on the gynaecological alert, its meaning and the surveys organised to stabilize the correlation between vaginal cancer and DES. DES is perceived as a risk factor associated with a high probability of adverse events in certain populations or under specific circumstances; everything possible must therefore be done to avoid these adverse effects and preserve the health of the population. In other words, not only must every 'reasonable precaution' be adopted, but each must also be implemented. The second discourse is pharmacological and focuses on the experiments conducted on mice and other animals to study the induction of tumours and the fate of DES within the body. It mobilizes statistical tools to draw conclusions regarding the best way of defining thresholds, in looking for zones of marginal effects (or better, of no effect) and conditions of relatively safe use. It aims at a controlled use of the substance under investigation. The third discourse focuses on animal physiology and nutrition. Also based on experiments with laboratory animals, it presents experiments with natural and artificial hormones as evidence of their identical fate and effects in the body. Nature and artifice being equivalent, the latter will not do more harm than the former. DES, as many other natural hormones, is rapidly eliminated from the body. Moreover, a consequence of the quasi-biological nature of DES action is that potency and carcinogenicity are two sides of the same coin. This physiological discourse is elaborated with nutrition studies focusing on the growth-enhancement effect of DES and its value for agricultural productivity. A more general agricultural-engineering perspective combines this veterinary expertise with economic considerations regarding the positive impact of abundant and cheap meat in balancing costs and benefits. In this perspective, the detection of residues is deemed important since traces of DES in edible meat initially went unnoticed; indeed, according to FDA critics, they were not even looked for. The technical answer consisted in seeking a new system of chemical measurement that would replace the mouse assay, gradually viewed as not sensitive enough. This turned out to be insufficient, however. If controlled use was to be maintained, sanitary inspection and legal action independent of DES producers and users would have to be reorganised in order to improve the follow-up on cases and assign responsibilities.

The FDA's position was dramatically affected by the dynamics of the controversy in general and of the hearings in particular: gradually the general trend in debate shifted from upholding controlled use to implementing a general ban. This gradual shift affected both facts and meanings; consequently, banning agricultural DES became scientifically, morally and administratively inevitable. In the spring of 1972, even the reluctant Bureau of Veterinary Medicine had rallied to the idea of a ban on some of the agricultural uses of DES:

> The principal source of estradiol and its metabolites in our foods could be from meat...Estradiol is a natural hormone in man, and the metabolic pathways are already established so that in the event any is absorbed, it will be rapidly metabolised and excreted. DES, on the other hand, is a synthetic estrogen having 10 times the potency of estradiol and is much more slowly metabolised and excreted...If violations of the 7-day withdrawal period should occur, the probability that muscle as well as kidney and liver, will be above what we now consider an 'acceptable' level would be real. (...) Until sufficient time has elapsed to clear the animal's metabolic pool of DES, we are dealing with a population of hyper-estrogenized animals that cannot be considered normal in this regard. We should like to have better evidence than we have now to conclude that 7 days are sufficient time to reduce the DES to the low physiological level of estrogenic activity normally found.[8]

Two months after writing this, Friedmann accepted a ban of DES premixes, with the FDA leaving ear implants as the only authorised form of DES administration in cattle. Although this shift did not formally abolish the 'controlled use' policy, it made it much more difficult to implement, first of all by attributing greater significance to the residue-detection issue and the mere presence of DES in meat. This became visible a few months later, when the Nader-campaign lawyers announced that: a) the U.S. Department of Agriculture (USDA) consumer service was finding DES residues in statistically significant numbers of carcasses (in less than one year – as the search was reinforced – the percentage of positives shifted from 0.5% to 2%); and b) they would sue the FDA for not having done anything about it. During the July 1972 hearings, public-health officials, as well as the director of the NCI, insisted on the existence of low-dose carcinogenesis and on the need for caution. Edwards and the head of veterinary medicine thus rediscovered the arguments of agricultural scientists, who had pressed in distinguishing between DES feeds – which should be prohibited – and DES implants, which were easier to control and required much lower doses than the feeds to obtain allegedly the same nutritional and agricultural benefits (US Congress, Senate, 1975, pp. 206–11). Reliance on the performance of a technically and politically reinforced detection apparatus was what finally triggered the ban of implants one year later, when it became public knowledge that the number of carcasses containing DES detected by the USDA surveillance system had escalated to 10% of the animals tested.

Both crises – in medical and agricultural DES – actually reinforced each other. The medical alert radically changed views on the enhancement of human reproduction through hormonal treatments, turning risks into proven adverse consequences. This had two effects on the agricultural scene and the discussion of animal enhancement. First, it gave very concrete meaning to the translation of animal modelling results to humans that had been at the core of all the arguments around the carcinogenic risks of the drug. If the correlation established in humans by Herbst and reinforced by the National Cancer Institute survey was accepted, this meant that DES given in therapeutic dosage was inducing tumours both in humans and in animals. Second, the medical crisis reinforced doubts about the policy the FDA had implemented since DES had been put on the market. Blind to the doubts raised by a number of physicians regarding the clinical value of the pregnancy indication, the agency had allied itself with a segment of the gynaecological elite and the DES-producing pharmaceutical firms. A similar scenario took place within the division of veterinary medicine, leading the agency to support the claims of nutrition scientists and feed producers.

Although specific connections between DES and risks have changed over the years, the geography of public expertise – with 'public health' on one side and 'industrial agricultural management' on the other – roughly matches the contrasting standpoints of those who pleaded for a complete ban on DES in farming and those who argued for a controlled use of the molecule. It also reflects the polarization of the debate between the 'Democrat-consumer alliance', including its allies among the scientists of the National Cancer Institute (NCI) and the 'agro-industrial alliance' led by the American Feed Manufacturers Association and the American Cattlemen's Association.

The statements of FDA experts and other government officials are however difficult to convey within the framework of regulatory capture, as they are contradictory and rapidly changing. As once remarked by Representative Fountain, the Chairman of the House Committee conducting the hearings, there was an underlying conflict of expertise between the U.S. Department of Agriculture and the US Public Health Service, but there were also deep divides within the FDA. Tensions were not only related to the form of expertise favoured within each branch of the administration, they also touched upon the moral economy of the agency, its conception of its role *vis-à-vis* 'the American public' and the acceptability of enhancement as a health-related objective. However, drawing from a broad understanding of these issues, FDA officials considered that humans and animals should not be conflated: whereas using drugs to improve human bodies not affected by a pathological situation was considered unacceptable and brandished as the FDA's main argument against mass 'off-label' prescription of DES, animal enhancement was never an issue in itself. It was problematic and remained a risk for human health, that is, a routine agricultural practice with unexpected adverse consequences for human bodies. During the entire hearings, not a single word was uttered on animal well-being.

The conjunction of a medical and an agricultural crisis, which gave new importance and meaning to the problems of animal enhancement – or rather of animal augmentation – was all the more powerful in that links between these two uses of DES were at work at various levels. First, there were technical links. Toxicological assessment of both usages relied on the same tools: assay procedures and modelling strategies. A mouse test was widely used to reveal the potency of compounds with oestrogenic effects (Gaudillière, 2005). The same animal model was also central to the study of carcinogenesis: all experiments suggesting that a given dosage of DES induced the formation of tumours, providing the sketchy 'dose-response' relationship then available, were first conducted on mice (Gaudillière, 2006).[9] Second, there were administrative and institutional links, since the FDA was the agency responsible for the regulation of drugs for both human and veterinary uses. FDA Commissioner Charles Edwards issued recommendations, comments and responses for both usages of DES, not always at the same time but in close connection. Even if two separate divisions were in charge of regulating drugs respectively for human medicine and veterinary products, the mere existence of a single administrative roof facilitated articulated choices. Given the history of the FDA, controlled use and pharmacological evaluation dominated the discussion in both segments. A peculiar consequence of this administrative unity was the nature of the correspondence the agency received. The hundreds of letters sent to the FDA after 1971 on the subject of DES rarely failed to mention humans and cattle simultaneously as the basis for their concerns.[10] Responses from the FDA initially tried to keep the regulatory issues apart, but this proved increasingly difficult as the controversy developed.[11] Third, the unity of DES was taken as an obvious fact in the public arena, notably by the media and consumer organisations. Most newspaper articles commenting on agricultural DES thus put in reminders of the carcinogenicity of therapeutic DES (although the opposite was rarely true).[12]

3 Food OR drug: French DES and the invisibility of animal enhancement

Choosing the French debate as a point of comparison for this 'food and drug' controversy is both relevant and problematic: relevant because the trajectory of DES uses in France is comparable to the American case even though the respective debates led to very different outcomes; problematic because agricultural DES never gained the kind of public visibility it had in the United States. Regulatory discussions took place behind the closed doors of the Ministry of Agriculture and not in the national assembly or other public arenas, even though the 'meat and hormones' issue repeatedly surfaced in the media.[13]

DES was barely visible in France before 1983, when daily newspapers began to document a 'medical scandal'. The affair originated not only in the 'unproven' use of DES in pregnant women, like in the United States, but also in the fact that gynaecological uses of DES were still current practice more than ten years after Herbst announced that this was a serious public health issue in the early 1970s. The medical commentator of the prestigious daily *Le Monde* – by no means a radical newspaper – reminded its readers that the single change that had occurred in French guidelines for medical uses of DES between 1971 and 1983 was the fact that the famous *Vidal* dictionary, a commercial handbook listing drugs, their composition, indications and side-effects, had modified the DES entry to include pregnancy among the contraindications of the drug (Escoffier-Lambiotte, 1983).[14]

In contrast to the United States, the issue was less often linked to a feminist critique of medicalisation: for instance, the link with the pill was never made, much less accepted in the French context. Medical DES was viewed instead as one more illustration of an insufficient control of the drug market. However, given the U.S. experience, one might have expected an impact of the medical affair on the 'meat and hormones' issue. What actually happened was just the opposite: the two problems remained unrelated, not only in the scientific literature, but also in mainstream media. Until the mid-1990s, not a single article about DES daughters mentioned the issue of oestrogens or oestrogen-like substances in animal feeds and meat.

A straightforward explanation for this difference could be that in France, the uses of DES in agriculture were limited. Using DES as growth enhancer had actually been explicitly forbidden since the early 1970s. This does not, however, warrant the conclusion that there was nothing to discuss: feeding animals with sex hormones was possible. The use of DES as an additive was not authorised, but DES and other oestrogens were veterinary drugs, meaning that professionals could prescribe them for any treatment they deemed necessary. In other words, feeds supplemented with hormones or antibiotics were not prepared and distributed, but veterinarians could consider that oestrogens played an important role in improving the health of animals or in preventing disease, and prescribe them on a rather large scale for animals that were not affected with any particular disease. Limitations in the use of DES had more to do with whether veterinarians prescribed them or not and with limited access to the substance, rather than with a legal ban. Hence a large industrialised farm that had good connections with agricultural technicians and veterinarians might not have found it too hard to obtain hormone treatment for its stock.

Occasional reports testified that the practice existed but was probably limited. In 1972, for instance, the national press echoed the legal action taken by the agricultural administration against a large-scale farmer and veterinarian from the Bordeaux area, who had implanted an entire herd (*Le Monde*, 1972). A more convincing testimony of the existence of the practice is the fact that in 1984, one year after the publication of the first epidemiological study documenting the existence of DES daughters in France, the Socialist government legalised the uses of sex hormones and their analogs as cattle growth-enhancers; the previous ban was thus replaced by a system of marketing permits and a preliminary evaluation of toxicity. DES was not on the list of authorised growth-stimulating substances published in 1985, but natural oestrogens and their analogs, whose carcinogenic potency had been discussed in parallel with DES for more than 25 years, were.

Before 1971, agricultural DES had been discussed only by veterinarians in a series of dissertations at the prestigious veterinary school of Alfort. These research reports focused on the benefits of the drug: accelerated growth, a more efficient use of nutrients, and by and large the value of DES for a more productive cattle husbandry. The main topic was therefore hormonal regulation and the ways in which DES could promote growth. Agricultural DES was no more visible in the farmer's press. The rare articles found in the widely circulated journal L'élevage only mentioned hormonal feeding when discussing U.S. agriculture.[15]

The issue of 'meat, hormones, and food additives' did materialize in the mainstream media but only in a vague and unspecific way. As of the late 1960s, the media occasionally raised questions about the consequences of industrial agriculture in terms of unsavoury food and of the potentially dangerous effects of food additives. The U.S. situation was often mentioned, both as an example of the excesses originating in the industrialisation of meat production and as a reference in cautious administrative regulation. One of the first articles of the kind alluded to the DES and chicken debate of the 1950s, presented the list of authorised additives and complained about the lack of regulation in France, explaining that:

> After having accepted the use of oestrogens in husbandry, for instance to accelerate the growth of chickens, it became clear that this practice was not without dangers for human health and was banned. The FDA had first authorised such treatment. A French bulletin for farmers thus claimed: 'Given the high scientific and moral authority of the US agency in charge of controlling the uses of hormonal meat-enhancers, we can be fully reassured.'
> (*Libération*, 1963)

The status of DES and hormone treatment in the media after the emergence of the U.S. crisis remained quite ambiguous. A common idea was that a problem of the magnitude that had been reached in the United States was impossible in France, where DES use was officially forbidden. However, concerns kept mounting. As the *Nouvel Observateur* explained in 1971: 'Feeding animals with hormones is forbidden in France but control is for the most part theoretical. In the United States, it is estimated that eight chickens out of ten are treated. And this does not take into account the cancerous animals Ralph Nader discovered'.

The question of control, fraud and inspection was indeed central. The main focus of regulation remained the slaughterhouses, where no animal treated with oestrogens could enter the human food chain; veterinary inspections led to specific measurements of DES but only when suspicion of fraud was strong enough to justify action. Articles addressing the 'meat and hormones' problem after 1971 did so in two different ways. First, they gave regular coverage to sanitary intervention and protests from veterinary inspectors who claimed they did not have enough manpower and the technical means to control the slaughtered cattle. The second type of article echoed consumer concerns about the quality of food. DES was discussed in articles on food additives and their regulation: they usually argued in favour of the establishment of a new infrastructure of public laboratories to test the toxicity of these scarcely known molecules. These concerns were also reported in the farmers' press, which had signalled a mounting problem with consumers' misinformed perception of the conditions necessary to modern agriculture and their misguided pleas for absolute safety.

Although we do not know much about the internal debates of experts and officials within the agricultural administration, it is clear that in the 1970s they felt it was politically necessary to reinforce regulation. In the spring of 1973, following a strike of meat inspectors, who had once again pushed the question of slaughterhouse control to the forefront, the Ministry of Agriculture issued a decree, specifying that hormones could only be administered under veterinary prescription, and that treated animals would be slaughtered only if they had a certificate specifying the motive, time and duration of hormonal injections. Moreover, if inspectors suspected that animals had been treated and sent to the slaughterhouse without a certificate from the prescribing veterinary, they were to be excluded from the food chain.[16] A comparison with the United States was again used as a reassuring factor. The minister at the time explained: 'Worries are rather exaggerated. Although many countries with a very high level of food hygiene, including the United States, still use oestrogens in animal feed, these substances are only authorised in France for therapeutic purposes, and inoculated by a veterinarian under very specific conditions...[the 1973 decree] protects the French consumer from the one single significant danger, namely – even if the question remains controversial – the residues of DES and other artificial oestrogens' (*L'élevage bovin*, 1974). A second step, taken in 1976, was the general prohibition of all uses of oestrogens and their analogs in veterinary medicine. The measure was taken in the name of safety after consumer organisations had threatened to boycott all veal. Both cattle breeder and veterinarian unions massively criticized the new law. As already mentioned, this mini 'beef and hormones' crisis did not leave vivid memories: the question of oestrogen carcinogenicity was barely mentioned when this general prohibition was later abolished, in 1984.

Given this low visibility, what evidence and what forms of expertise were legitimate, acceptable and actually used when discussing DES as a tool for animal enhancement in France? Although it was present after 1983, the public-health discourse focusing on DES-induced human cancers did not connect with the agricultural debate, and the pharmacological discourse focusing on thresholds and dose-response patterns played a rather limited role in the discussion. Consumer safety was actually seen in terms of: a) preventing fraud, with an improved system of meat inspection; and b) opposing the use of artificial additives and physiological or natural regulators. The distinction between natural and artificial hormones brought about the idea that DES posed a special problem since it was not a hormone but a synthetic analog, structurally unrelated to sex steroids. This perspective was essential in the design and application of the 1984 law: the marketing permits granted on that basis excluded all products but natural sex hormones (oestrogens, progesterone and testosterone).

The reference to the US controversy, albeit discreet, nonetheless induced a new wave of research on DES effects, this time both physiological and toxic. Researchers at the Alfort veterinary school launched a series of nutrition experiments focusing on the effects of small DES doses on laboratory rats. One key point was that natural hormones could act as carcinogens, but at much higher concentrations than DES and, more importantly, they were rapidly metabolized and eliminated. The physiological, rather than pharmacological, culture of the French veterinarians resulted in a specific modelling of DES effects. The assay designed by R. Ferrando and his colleagues at the Alfort veterinary school did not measure the toxicity of specific quantities of DES introduced into the meat or food given to the test animals; instead, it measured the effects of the product that might enter human nutrition, that is, the meat from cattle supplemented with DES. The plea for taking into account this kind of experimentation was that the animal body processed DES in unknown ways; thus indirect exposure to DES might account for a much less dangerous level of risk than direct exposure or, alternatively, newly synthesized metabolites and residues – as seemed to be the case with DES – might on the contrary aggravate the risks:

> Animal organisms provide for a 'metabolic relay' between the feeding rations consumed by wild or domesticated animals and humans who use parts of these animal bodies for its daily nutrition. Although decisive, the role of this relay has been neglected if not completely ignored. It is indispensable to assess the toxic effects of chemical substances used either to increase the performance of farm animals or to protect them from the negative consequences of endemic infections. Current studies on laboratory animals paradoxically reinforce the safety of farms rather than that of consumers. The existence of a 'relay' can change everything. (Truhaut and Ferrando, 1976, p. 422)

The technology of 'relay toxicity' actually convinced the Alfort scientists that DES was a special case, not to be equated with other (natural) hormones:

> From the point of view of food hygiene, artificial oestrogens and sex hormones show radical differences. While the latter are rapidly metabolized and eliminated from the organism, artificial substances leave 'residues' that persist in meat and other animal products. Such residues rightly attract the attention of hygienists since the consumer's organism cannot destroy them and (as is the case with DES) eventually create specific dangers. (Vuillaume, 1975, p. 5)

Given the absence of public expertise, this level of proof and form of legitimacy proved decisive to the French Ministry of Agriculture where veterinarians occupied a central position not only as (sole) experts of animal husbandry but also as organisers and controllers of meat inspection. Arguing for a lifting of the 1976 ban, R. Ferrando (1980, p. 568) thus explained:

> The experiments of relay toxicity show that natural steroids – male or female – as well as their derivatives are without danger for the consumer. Their use facilitates farming. It helps to prevent diseases and multiple therapeutic interventions, which are both costly and polluting. In contrast, the use of a non-steroid anabolic drug like DES raises major problems.[17]

Referring to the situation created by the 1976 ban, his colleague R. Vuillaume (1983, p. 7) felt that the new expertise had provided powerful means of avoiding health risks while providing farmers with legitimate means of increasing their productivity:

Many farmers could not abandon a practice essential to the profitability of their operations. As veterinarians refused to give them products whose use had become illegal, they accepted the offers of quacks who sold – and often administered – fraudulent products. While veterinarians had always been using substances they knew to be innocuous to the consumer, these quacks did not hesitate to spread artificial chemicals (DES, hexstrol) they could buy inexpensively and administer by way of intramuscular inoculation invisible to the slaughterhouse inspector, without any concern for the fact that such chemicals left important residues at the point of injection, thus creating a very serious health risk.

Taking into account the contrast between the United States and France in their respective approach to public expertise, one could interpret the French dissociation of the two uses of DES as the manifestation and consequence of three features.

The first and more general feature was the prevailing role played by the separate professional regulation of meat and drugs. DES in its two different uses was handled by two groups of experts that interacted rarely and had substantial autonomy. This delegation of expertise to their respective professional bodies, veterinarians and physicians, was all the more important in the 1970s as it was not offset by strong industrial or administrative practices of surveillance and control of potentially dangerous chemicals. Thus, the contrast between the conflation of issues related to food and drugs in the United States and their separate treatment in France was strongly reinforced by the dramatic differences in mandate, means and mode of intervention that characterized the FDA on the one hand and the French Ministry of Agriculture on the other.

A second feature associated with this professional separation in addressing DES use was the absence of a shared 'regulatory science'. In France, the questions raised by the use of agricultural DES were not addressed with the pharmacological and toxicological models commonly used in the regulation of therapeutic agents in human medicine. Risk was evaluated with experimental models built on the physiological traditions of academic veterinarians with few or no industrial connections.

Third, if the shortcomings of attempts to enhance human reproduction were definitely present in the French debate on the medical uses of DES, through the DES victims' critique of the unnecessary and dangerous treatment of pregnancies, discourse on risk was confined to the medical sphere. Whatever problems might arise for humans or for animals from enhancing livestock, that is, the alteration of animal qualities for the purpose of human consumption, these were approached exclusively from the perspective of food quality and framed in terms of the use of natural versus artificial hormones. Consequently, the accepted but narrow professional definition of risks excluded low-dose carcinogenesis, which had emerged as central boundary category in the handling of U.S. public expertise.

4 Conclusion

This chapter has examined contrasting approaches in addressing the questions raised by the administration of DES to animals for food production, within the context of the DES medical affair and ensuing public debates on the regulation of sex hormone prescription. In the American case, the problems raised by the use of this molecule, in agriculture and in medicine, were all handled at the same time, as part of a single process of condemnation and regulation. In France, DES became a public issue only in the late 1980s, and only as a medical issue, when studies of DES daughters were initiated and patient activism emerged. No link was ever made between the medical and the agricultural uses of DES, and the (putative) circulation of DES in meat, which had been at the centre of FDA action in the United States, was never raised.

This contrast may be understood on the basis of differing medical cultures, ways of regulating drugs and targets of social mobilization. In this latter respect, historians have emphasised the role played in the United States by social movements, first and foremost environmentalism and feminism, and the visibility these movements gave to problems of low-dose carcinogenesis. Analogous movements in France existed, but focused on other issues such as nuclear energy and abortion; the regulation of DES thus remained, at least before the 1990s, a professional matter delegated to the respective jurisdictions of physicians and of veterinarians.

The politics of consumption visible in the American 'meat and DES' crisis was indeed a major feature of the US post-war culture. This helps us to understand why the issues raised by DES uses were framed in terms of problematic practices in agriculture with increasing risks to human health. Unlike debates on the quality of drugs, in which the suffering and dependency associated with diseases are essential, debates on food production focused not only on questions of its safety for human consumption but also on the legitimacy of handling animals in an industrial manner to foster quantity and cheapness. Within the perspective of consumer movements, buyers needed to be organised in order to counteract the power of industrial monopolies and gain influence in the regulatory arenas. However, the main focus of their action remained the final user and his or her ability to make informed choices in order to improve their way of living. Fair trade was a key concept in this context; practically speaking, investments were made to set up information campaigns, prepare judicial actions to claim compensation in case of damage, devise tests to measure risk and undertake cost-benefit analysis. As illustrated in the congressional hearings, the latter tools proved powerful instruments for bringing together heterogeneous social, economic and professional interests under the umbrella of a human-centred assessment of the limitations of industrial agriculture.

In contrast to the situation in the United States, when the uses of sex hormones in agriculture were discussed in France, before the emergence of a medical crisis, the central position of veterinarians in the regulatory process did not provide them with a sufficiently strong basis to challenge the practices of the medical profession and thus overcome the barrier that set medicine and agriculture apart as distinct social worlds. Moreover, the French debates were less concerned with the industrial nature of modern agriculture than with the problems of food quality and tastiness; animal manipulation for productive purposes remained unproblematic unless it involved deviant practices. Finally, the public debate on the problems of feeding animals with hormones was framed in terms of synthetic chemicals versus natural substances: DES was considered harmful because of its artificial, non-physiological, origins. This left open the option of using 'natural' hormones as a potent alternative to DES, without much concern as to their carcinogenicity for both humans and animals.

Despite the differing trajectories of DES use in agriculture that emerge from these two configurations, its treatment in both countries left out all issues related to animal welfare, revealing a common trait to the visions of animal and human enhancement that were at stake in these debates. In both contexts, debates on the uses of DES relied on a differential understanding of the term enhancement, according to whether it was being applied to humans or to animals. The condemnation of DES as a means of medicalising pregnancy by managing situations of uncertain risk rather than actual disease mobilized the idea that DES prescription had been a risky, unnecessary and in the end illegitimate attempt to improve human bodies. In contrast, what was ultimately at stake in the debates on agricultural DES was not enhancement in the sense of improvement of the animal for its own sake, but enhancement in the sense of augmentation of bodily capabilities for the sake of human meat consumption. In both countries, the use of DES in agriculture was banned fundamentally because of the side-effects this practice could have on humans, with no concern for its impact on animals. This could imply that there is simply no such thing as animal enhancement since the concept, as it is used today, should designate the improvement of animal life and not their augmented productive capacity.

There are however two arguments in favour of not restricting the meaning of animal enhancement to this narrow definition of an improvement benefiting the enhanced individual. The first argument is pragmatic: the notion of 'growth enhancement', with its emphasis on the augmentation of animal properties, was actually used in debates about agricultural DES, at least in the United States. The second argument is analytical. Viewed from a practical point of view, the risks associated with such practices of bodily modification both in animals and in humans, that is, whether their aim be improvement or augmentation, are connected in many ways. The DES trajectory accordingly teaches us, first of all, that agriculture and medicine are not two worlds apart: the same tools, techniques and sometimes personnel circulate between them, thus making it possible to enhance both types of bodies. Second, it illustrates how enhancement practices are developed in both agriculture and medicine, at the boundary between the normal and the pathological, where forms of treatment are turned into attempts at improvement. This extension of the therapeutic realm, on the basis of interventions whose legitimacy were often called into question, was a key factor in generating new risks for human and also for animal health. In other words, despite differences in local 'ways of regulating' drugs and of generating expertise in France and in the United States, the questions raised by the practices of animal augmentation may ultimately be highly relevant in evaluating our most recent attempts to improve humans.

Notes

1   'DES Livestock Implants Are Prohibited by FDA because of Cancer Link', *Wall Street Journal*, 26 April 1973.

2   This section is an amended version of an excerpt, 'Agricultural DES in the United States', from a previous publication entitled 'Food, Drug and Consumer Regulation: The 'Meat, DES and Cancer' Debates in the United States' (Gaudillière, 2010). The author wishes to thank Pickering and Chatto for permission to reuse this material. On the early history of DES, see: Bell (1980); Gaudillière (2006).

3   For the expertise conducted during this first phase, see the National Academy of Science Report (1959).

4   For more on this aspect, see, in addition to Marcus's book, Dunlap (1981); Hays (1987); Proctor (1985); Gillespie *et al.* (1984); Brickman *et al.* (1985).

5   On the 1962 Act in general, see Temin (1980); Daemmrich (2004). On the amended clause, see *Congressional Record*, 87th Congress, II, p. 12713.

6   In addition to the aforementioned literature on the DES medical crisis, see also: Bonah and Gaudillière (2007).

7   FDA archives, NDA 4038-DES Lilly, Report of the National Academy of Sciences, 1969.

8   L. Friedman to V.O. Wodicka, Director Bureau of Foods, 'DES', 8 February 1971, in US Congress, Committee on Government Operations, House of Representatives, 1972, pp. 174–5.

9   Animal models constitute another form of animal enhancement that implies changing animal bodies, in order to bridge the gap between human and animal bodies and thus increase the capacity of animals to 'represent' the course of (human) pathologies. This third meaning of enhancement, at stake in the DES trajectory, is beyond the scope of this chapter.

10 Many of these letters emanated from citizens with no apparent engagement in consumer or environmental groups. A certain Mrs. Hutchinson from Greensborough (NC) thus simply wrote in September 1972: 'I appreciate your concern about Hexachlorophene by taking it off the market. But why overlook DES? I realize you have taken it off the market to animals, but why leave it for human consumption?' National Archives, RG 88, Box 4662.

11 Answering a certain Ronald Riba from Arlington, Illinois, about the fate of the drug, the Bureau of Drugs and the Bureau of Veterinary Medicine, for instance, wrote separate answers, respectively, focusing on the redefinition of therapeutic indications and on the imminent withdrawal of DES implants in application of the Delaney clause. Both letters nonetheless needed to refer to the interventions of the other bureau. National Archives, RG 88, Box 4876.

12 One unexpected text of this sort was a devastating article by the editorialist of *Science*, Nicolas Wade, written in 1972.

13 As a consequence, sources are more difficult to find. As administrative archives are not (yet?) accessible, the following account relies on analyses published in the press, including medical and agricultural scientific journals (e.g., *Annales de zootechnie*), professional journals of farmers' and breeders'

organizations (e.g., L'élevage) and mainstream daily and weekly papers (e.g., Le Monde).
14 For an analysis of the French medical crisis, see Fillion and Torny (forthcoming 2016).
15 'Production industrielle de viande bovine aux USA', L'élevage, n° 1, (1973): p. 3.
16 This issue was addressed over several months in the professional journal L'élevage.
17 See also his 'Introduction', 1983, p. 12.

References

———(1956) 'Hormones and Cancer', *New York Times*, 29 January.
———(1958) 'Bill on Food Additives Gains', *New York Times*, 14 August.
———(1963) 'Du poison dans votre assiette', *Libération*, 11 December.
———(1971) 'Les poisons de la table', *Le Nouvel Observateur*, 29 November.
———(1972) 'Un éleveur et un vétérinaire sont inculpés pour utilisation illégale de produits estrogènes', *Le Monde*, 27 November.
———(1973) 'Production industrielle de viande bovine aux USA', *L'élevage*, n°1: p. 3.
———(1973) 'DES Livestock Implants Are Prohibited by FDA Because of Cancer Link', *Wall Street Journal*, 26 April.
———(1974) "Des craintes exagérées", *L'élevage bovin*, n° 7: p. 4.
Apfel R.J. and Fisher S.M., (1984) *To Do No Harm, DES and the Dilemmas of Modern Medicine* (New Haven: Yale University Press).
Bell S. (1980) *The Synthetic Compound DES, 1938–1941: The Social Construction of a Medical Treatment* (PhD dissertation, Brandeis University).
Bonah C. and Gaudillière J.P. (2007) 'Faute, accident ou risque iatrogène? La régulation des évènements indésirables du médicament à l'aune des affaires Stalinon et Distilbène' *Revue Française des Affaires Sociales,* n 3–4, 123–51.
Brickman R., Jasanoff S. and Ilgen T. (1985) *Controlling Chemicals: The Politics of Regulation in Europe and in the United States* (Ithaca: Cornell University Press).
Carson, R. (1962) *Silent Spring* (Boston: Houghton Mifflin).
Daemmrich, A. (2004) *Pharmacopolitics: Drug Regulation in the United States and Germany* (Chapel Hill: University of North Carolina Press).

Dieckmann W.J., Davis M.E., Rynkiewicz L.M. and Pottinger R.E. (1953) 'Does the administration of diethylstilbestrol during pregnancy have therapeutic value?', *American Journal of Obstetrics and Gynecology*, 66: 1062–81.

Dunlap T.R. (1981) *DDT: Scientists, Citizens, and Public Policy* (Princeton: Princeton University Press).

Dutton D.B. (1988) *Worse Than Disease: Pitfalls of Medical Progress* (Cambridge: Cambridge University Press).

Escoffier-Lambiotte C. (1983) 'Une monumentale erreur médicale. Les enfants du distilbène', *Le Monde*, 16 February.

Ferrando R. (1980) 'Les anabolisants stéroïdiques et non-stéroïdiques et l'élevage', *Bulletin de l'Académie Nationale de Médecine*, 164: 568–72.

Ferrando R. (1983) 'Introduction' in *Anabolisants en production animale*, International Symposium, Paris, 15–17 February: p. 12.

Fillion E. and Torny D. (2016) 'Un précédent manqué: le Distilbène et les perturbateurs endocriniens. Contribution à une sociologie de l'ignorance.' *Sciences sociales et Santé*, spécial: *Perturbateurs endocriniens* (forthcoming).

Gaudillière J.-P. (2005) 'Better prepared than synthesized: Adolf Butenandt, Schering AG and the transformation of sex steroids into drugs', *Studies in History and Philosophy of the Biological and the Biomedical Sciences*, 36: 612–44.

Gaudillière J.-P. (2006) 'Hormones at risk: cancer and the medical uses of industrially produced sex steroids in Germany, 1930–1960', in Schlicht T. and Ströhler U. (eds.), *Risk and Safety in Medical Innovation* (London: Routledge), pp. 148–69.

Gaudillière J.-P. (2010) 'Food, drug and consumer regulation: the "meat, DES and cancer" debates in the United States', in Cantor D., Bonah C. and Dörries M., *Meat, Medicine and Human Health in the Twentieth Century* (London: Pickering and Chatto), pp. 179–202.

Gillespie B. Eva D. and Johnson R. (1984) 'Carcinogenic risk assessment in the United States and Great Britain: the case of Aldrin/Dieldrin', *Social Studies of Science*, 14: 265–301.

Greenwald P. Barlow J.J., Nasca P.J. and Burnett W.S. (1971) 'Vaginal cancer after maternal treatment with synthetic estrogens', *New England Journal of Medicine*, 285: 390–2.

Hays S.P. (1987) *Beauty, Health and Permanence: Environmental Politics in the United States, 1955–1985*, (Cambridge: Cambridge University Press.

Herbst A.L., Ulefelder H. and Poskanzer D.C. (1971) 'Adenocarcinoma of the vagina: association of maternal stilbestrol therapy with tumor appearance in young women', *New England Journal of Medicine*, 284: 878–81.

Longgood W. (1960) *The Poisons in Your Food* (New York: Simon and Schuster).

Marcus A.I. (1986) *Cancer from Beef. DES, Federal Food Regulation and Consumer Confidence* (Baltimore: Johns Hopkins University Press).

Marks L. (2001) *Sexual Chemistry. A History of the Contraceptive Pill* (New Haven: Yale University Press).

Meyers R. (1986) *DES, The Bitter Pill* (New York: Putnam).

Morgen S. (2002) *In Our Own Hands: The Women's Health Movement in the United States* (New Brunswick: Rutgers University Press).

Nader R. (1965) *Unsafe at Any Speed: The Designed-in Dangers of the American Automobile* (New York: Grossman Publishers).

National Academy of Science Committee on Animal Nutrition, Subcommittee on Hormones (1959) *Hormonal Relationship and Applications in the Production of Meats, Milk, and Eggs* (Washington DC: National Academy of Science).

Pfeffer N. (1992) 'Lessons from History: The Salutary Tale of Stilboestrol' in Alderson, P. (ed.) *Consent to Health Treatment and Research: Differing Perspectives* (London: Social Science Research Unit).

Proctor R. (1985) *Cancer Wars* (New York: Basic Books).

Rifkin J. (1992) *Beyond Beef: The Rise and Fall of the Cattle Culture* (New York: Dutton).

Seaman B. (1969) *The Doctors' Case against the Pill* (New York: P.H. Wyden).

Shell O. (1984) *Modern Meat* (New York: Random House).

Silber N. (1983) *Test and Protest: The Influence of Consumers Union* (New York: Random House).

Temin P. (1980) *Taking Your Medicine: Drug Regulation in the United States* (Cambridge: Harvard University Press).

Truhaut R. and Ferrando, R. (1976) 'Résultats de huit ans de recherches sur l'évaluation toxicologique des additives à l'alimentation animale par la méthode de la toxicité relais'. *European Journal of Toxicology*, 9: 413–22.

Turner J.S. (1970) *The Chemical Feast: Ralph Nader's Study Group Report on the Food and Drug Administration* (New York: Grossman Publishers).

US Congress (1958) 'An Act to Protect the Public Health by Amending the Federal Food, Drug, and Cosmetic Act to Prohibit the Use in

Food of Additives Which Have Not Been Adequately Tested to Establish Their Safety', *US Statutes*, pp. 1784–9.

US Congress (1962) *Congressional Record* 87th Congress II, p. 12713.

US Congress, House of Representatives, Committee on Government Operations, Hearings held on 11 November 1971, *Regulation of DES*, Testimony of C. Edwards, Commissioner FDA, Washington DC, Government Printing Office, 1972, pp. 49–54 and 76–84.

US Congress, Senate, Committee on Labor and Public Welfare, Hearings held on 20 July 1972, Regulation of DES, Testimony of C. Edwards, FDA Commissioner, Washington DC, Government Printing Office, 1975, pp. 206–211.

Vuillaume, R. (1975) 'Des viandes aux hormones', *L'élevage*, n° 3: p. 8.

Vuillaume, R. (1983) 'Oui aux anabolisants sans risque', *L'élevage bovin*, n° 12: p. 7.

Wade, N. (1972) 'DES. A Case Study of Regulatory Abdication', *Science*, 177: 335–7.

5
So Different and Yet So Similar: Comparing the Enhancement of Human and Animal Bodies in French Law

Sonia Desmoulin-Canselier

Abstract: *At first sight, the question of enhancement receives a totally different answer for humans and animals. Animal improvement is legally organized and institutionalized, through artificial reproduction, selection and biotechnologies, whereas human enhancement seems to be condemned outright. However, some elements do not fit into this clear picture. Numerous techniques used in breeding are now proposed for use in human reproduction. In doping, the comparison may lead to the counter-intuitive acknowledgement that animals are not always less protected than humans. Human embryos are now used for scientific research and for the long-term improvement of the human species. Some beloved animals may benefit from cutting-edge veterinary techniques. These various elements shed light on the blurred borderline between care and improvement in both human and animal bodies.*

Bateman, Simone, Jean Gayon, Sylvie Allouche, Jérôme Goffette and Michela Marzano, eds. *Inquiring into Animal Enhancement: Model or Countermodel of Human Enhancement?* Basingstoke: Palgrave Macmillan, 2015. DOI: 10.1057/9781137542472.0010.

Is there anything to be learned about human enhancement by comparing it to the age-old practices of animal enhancement? One possible approach to this question is to compare the way the law handles practices that can be considered forms of enhancements, as they are applied in animals and in humans. Indeed, studying the legal solutions adopted to regulate scientific techniques of intervention on human and animal bodies is of undeniable interest as a basis for reflection: law is a (supposedly) stable 'material' because it is mainly textual, and it is supposed to reflect the choices of a given society at a given time. However, a comparison of the legal approach to 'enhancement', as applied to humans and to animals, must first begin with an examination of two terms: enhancement and comparison.

None of these words are in the French legal glossary. They have no official specific meaning in the French legal sphere, at least for the moment. Nevertheless, this does not mean that the law has nothing to say. Common language may be sufficient to capture their significance. Here, the difficulty – and the interest – comes from the complexity and the polysemy of the terms. The word 'comparison' may seem easier to define, even though it encompasses numerous levels of meaning. 'Comparison' emphasises both similarities and differences, and implies an acceptance of the fact that the objects of analysis can in fact be placed side by side. The richness of this concept can be appreciated when bearing in mind that it is frequently associated with the image, the metaphor or the antithesis (Cassin, 2004, v° comparaison). In trying to explicitly state the reasoning behind this study, the term 'comparison' reveals the range of underlying questions. What, in actual fact, should be compared as objects of legal rules? The techniques that are used? The results that are sought? Should one be content with discussing only physical interventions or should one also explore the moral implications of transposing to humans what is used on animals (or *vice versa*)? These questions are all the more delicate in that they assume we know what 'enhancement' means, even though the term appears to be as polysemic as it is problematic. It is easy to understand that making someone or something 'better' does not involve the same action, depending on whether a moral criterion or a reference to physical performance is applied to evaluate the state of 'being better' (improvement). Determining the objective to be attained and the evaluator are both decisive measures, and yet both are undecidable if a scale of values is not presupposed. When showing an interest in biotechnologies and interventions on human and animal bodies, this becomes flagrant. The idea of 'enhancing' does not refer to the same concepts or to the same realities for animals and humans and, in consequence, does not lead to the same conclusions, depending on the viewpoint adopted. It should also be said that increasing physical or mental capacities, or adding new capacities, can concern individuals or their offspring, and through the latter, the group or species to which the individual belongs. This broadens the field of investigation even further.

At this stage, the author has to admit that all the questions raised will not find an answer. In a modest posture, I will propose some preliminary remarks, essentially taken from the French legal texts applying to techniques that might be considered as 'enhancement techniques'. The examples I have chosen do not constitute an exhaustive list, but they will hopefully contribute to shed new light on two related questions (or maybe a double question) – that of the comparison between humans and animals, and that of the ambiguous status of enhancement in our society.

Of what particular interest is it to start this broad questioning by studying the solutions adopted by French law? Various reasons could be put forward but there are two main reasons. The first concerns the position adopted by French law on the organisation of two fundamental and mutually exclusive categories: 'persons' (all legal persons including individual human beings, through the subcategory of 'physical persons', as well as human beings gathered within associations or companies, designated as 'moral persons') and 'things' (all legal things including animals). The second concerns the legal positions adopted with respect to practices subsumed under the term 'bioethics'. French law took a strong stand at quite an early stage on the use of biotechnologies, whether applied to animals or humans. The first legal texts concerning 'genetic enhancement' in animal breeding go back to 1966. As for the French laws on 'bioethics', they have attracted attention even beyond the French borders ever since they were first voted in 1994, and successive amendments to these laws have aroused considerable interest. At a different level, it can be noted that Descartes, the eulogist of the contrast between humans and animals, has left a strong imprint on French culture, although he is also an ambiguous figure when examining the implications of an intervention on the body, whether human or animal. Mind-body dualism lies at the heart of the Cartesian opposition between humans and animals, but it may also be the foundation of a reifying vision of the human body.

Thus, French law makes a clear distinction between the status granted to human beings and to animals as objects of law. The question of 'enhancement', as exemplified by practices such as artificial insemination or genetic selection, can be used to illustrate this point (I). A few recent developments have nevertheless revealed tensions in the way this distinction is structured, betraying possible flaws in the discourse affirming human specificity with regard to enhancement techniques (II). Legal innovations, such as those intended to protect the 'human species' from certain biotechnological applications, present ambiguities that should be emphasised if one wishes to have a better idea of how effective such innovations might be (III).

1 The law affirms the difference between humans and animals with regard to enhancement techniques

At first sight, the question of 'enhancement' is merely a new example of the traditional opposition in French law between rules applicable to humans and those concerning animals. According to the jurist and legal sociologist Jean Carbonnier (2004, p. 381), '[T]he human body is the *substratum* of the (human) person'. Although this principle has been undermined by recent law, it remains valid for the time during which the physical person is alive because, as the same author points out:

> [S]ince the will never appears to us unless it is linked to a body, it is not unreasonable to submit as a principle that it is the human body that makes the person. This is the meaning that can be attributed to the first law on bioethics (94-653 of 29 July 1994), in the sequence (art. 16 to 16-9) that it inserted in the Civil Code (except if we hesitate on the philosophy that inspired it): the primacy of the person is proclaimed in the frontispiece but in the text that follows it is amalgamated with respect for the body.[1]

It is, indeed, in a chapter of the French *Civil Code* devoted to 'respect for the human body' that the first law on bioethics incorporated the articles affirming, notably, that 'the law ensures the primacy of the person' (Article 16), that 'each person has the right to respect for their body' and that 'the human body is inviolable' (Article 16-1), thus intermingling considerations about protection of the physical person with those relating to protection of the human body. Conversely, the bodies of animals are placed under the aegis of the law on property (things that are either appropriated or can be appropriated), without birth or death modifying this reality in any way. It is therefore a principle of free disposal by the owner that presides over their destiny, even though exceptions and restrictions to this principle have multiplied to the extent that certain authors now challenge the notion that animals belong to the category of possessions (see in particular Marguénaud, 1998, Chronique, p. 205; Antoine, 2005). The legal definition of animals as 'sensitive living beings' (in the *Rural Code* since 1976, in the *Civil Code* since 2015) does not modify this basic legal situation. In this context, the question of 'enhancement' arises initially in a very different manner, according to whether it concerns human or animal bodies. Whereas, in the first case, it is tied to the problem of expressing one's will and respecting the consent of physical persons to undergo an intervention for the purpose of increasing their physical or mental capacities, in the second case, it refers to the problem of human power over sensitive beings, through the delimitation of rights exercised by an owner (or possessor). Other contributors to this book have, with good reason, insisted on focusing on this aspect.[2]

The laws covering recourse to reproductive techniques are part of this reasoning. The legal framework for 'medically assisted procreation' (Articles L. 2141-1 and following, *Public Health Code*), a term that encompasses all human assisted reproductive technologies in France, regulates these practices as medical treatment that responds to a 'parental project' (*projet parental*); this specific interpretation is granted to the human person and to the wishes expressed by him or her. The notion of 'public order' (*ordre public*: in other words a legal version of moral values accepted at a given time) is invoked to add legal constraints that limit these measures to persons capable of consent with parental projects deemed 'acceptable' according to the terms of the law. In other words, French law appears to affirm that reproductive technology is only a question of treatment and not of enhancement. Admittedly, the legislator recognises that these techniques 'permit procreation outside the natural process' (expression used in Law n° 2004-800, art. 24-I, 1°, modified by Law n° 2011-814, art. 28 & 31) but it is, first, a matter of 'curing an infertility' of a pathological nature that has been 'medically diagnosed', or of avoiding the 'transmission to the child or to a member of the couple of a particularly serious disease' (Article L. 2141-2). On the other hand, the *Rural Code* regulates 'the reproduction and genetic improvement of animals for breeding purposes' (Articles L. 653-1 and following, *Rural Code*). The objective of this law is clearly identified as the 'improvement' (so close to 'enhancement') (*amélioration*) of the quality of the equine, asinine, bovine, ovine, caprine and porcine species, rabbits, poultry and aquaculture species, as well as domestic carnivores (Article L. 653-1) with the economic goal of a more efficient management of 'animal production'. To achieve this, the *Rural Code* makes provisions for recourse to the operation of '*monte publique artificielle*' (public artificial breeding), covering several techniques of animal reproduction (including artificial insemination) that require artificial transport of genetic material outside its place of production, to ensure a quasi-public (delegated) service of assisted reproductive techniques (see, in particular, Desmoulin, 2006a, §192 and following).

It is difficult, therefore, to imagine more widely differing regimes than those for artificial human and animal reproduction, even though the techniques used are comparable and are the outcome of the same research.[3] The letter of the legal texts could even lead to the removal of medically assisted procreation from enhancement techniques if it is forgotten, too hastily, that the expected results are, in fact, not so different. Granted, there is no question of large-scale systemisation and planning as in the case of animals. However, medically assisted procreation is not exempt from any intention to enhance. As Simone Bateman (2004, p. 393) has summarised, the objective is either to 'overcome a natural dysfunction, which need not be accepted as inevitable' or to 'circumvent an obstacle to a pregnancy that is "naturally" possible and desirable'. In any event, the objective is to surpass the physiological limitations of the bodies – in the case of defective reproductive capacities or the lack of such capacities by recourse to 'Intracytoplasmic Sperm Injection' techniques and embryo transfers – even if doing so may provoke multiple pregnancies. Public health problems – such as the professional competence of the persons intervening in the removal, storage or transfer of embryos or gametes – have also led in both cases to public authorities strictly regulating such practices. However, establishing a parallel between practices that control human and animal reproduction seems to be eminently provocative (Fagot-Largeault, 2011).

This intellectual approach sheds light on reciprocal influences. One striking example is cloning. Unsurprisingly, since cloning is an artificial reproductive technique, we find in French law the usual opposition between humans and animals. Animal cloning based on somatic cell transfer was inaugurated with the birth of the ewe Dolly in 1996. It was expected that once implementation difficulties were resolved, it would be integrated into routine breeding practices, and French law provided the framework for dealing with this technique of '*monte publique artificielle*' (public artificial breeding) (Article L. 653-2 of the *Rural Code*[4]; Decree n° 2007-818 of 11 May 2007[5]). France, unlike other countries ('Clonage: le gouvernement néerlandais interdit le clonage de bovins', *Le Monde*, 1998), has never considered banning animal cloning. The public authorities saw in it a new way of making progress in 'animal production' (milk, meat and so on) by increasing the number of offspring of the animals with the highest performance. Conversely, human cloning was forbidden in all its forms and whatever the final goals pursued (Articles 214-2, 511-1-2, 511-18-1 of the *Penal Code*). As for reproductive cloning, it is criminally sanctioned as a 'crime against the human species'. First developed in the field of animal reproduction, this technique aroused such great fears that it would ultimately be applied to humans that specific legal measures imposing the most severe sanctions in the penal system were voted even before experiments were initiated.

This reaction caused such a commotion that animal cloning in turn became a subject of mistrust. French research activities carried out in public laboratories, very much at the forefront of animal cloning during the first years of the millennium, were gradually abandoned or suspended. The French Agency for Food Security (AFSSA, 2005), followed by the European Food Safety Authority (EFSA, 2008; 2009; 2012), submitted lukewarm views on resorting to animal cloning for breeding purposes. However, far from being merely anecdotal arguments in views expressed mainly concerning the possible toxicity of food, the consequences of adopting this technique on the life, health and 'welfare' of animals occupied a strong position among the arguments that ultimately led to the adoption of an extremely qualified point of view as to the interest of authorising the marketing of food products obtained from cloned animals. Thus, animal cloning has been suspended or banned (for the moment), not essentially for sanitary reasons, but because of the consequences for the offspring in terms of health and welfare. Analysing this argumentation, it appears that animals are no longer pure raw material: they are living entities and we have to take their health and welfare into account. At this point, one can see that reciprocal influences between the issue of human cloning and the issue of animal cloning exist not only in their respective practices, because of a common technique used, but also in the way questions are thought and arguments are built. Moreover, these influences are not limited to this point. Scientists who have developed an expertise in the field of cloning have pointed out that this technique may lead to new knowledge and inventions in the medical field. Recent parliamentary debates held between January and July 2011 in view of amending the 2004 law on bioethics (*Loi n° 2004-800 du 6 août 2004*) demonstrated that this argument had not become dead letter, since one of the proposed amendments concerned a restricted authorisation of somatic cell nuclear transfer, often referred to in public debates as 'therapeutic human cloning'. This technique, based on the transfer of the nucleus of an adult somatic cell into an enucleated oocyte to obtain totipotent cells and establish cell lines that will hopefully counteract physical deficiencies, lies at the heart of discussions concerning the borderline between treatment and 'enhancement'. The proposal was rejected, but the question remains, since the law – *Loi n° 2011-814 du 7 juillet 2011 relative à la bioéthique* – will have to be reviewed again in seven years.

The facts – in this instance, the physiological proximity of human and animal bodies and the similarity of the techniques used on both – may therefore weaken the affirmation of the principle that there is a fundamental difference in law between humans and animals. Law evolves, and in so doing, encounters tensions that could turn out to be flaws.

2 The law reveals weaknesses in the discourse affirming human specificity with regard to enhancement techniques

French legal rules usually indicate that a clear distinction has been maintained between solutions related to humans and to animals. However, a closer examination of these rules reveals developments that have opened breaches in the position asserting human specificity with regard to 'enhancement' projects. This can be verified thanks to a dual tendency in French law. On the one hand, recent changes affecting the status of the human body tend to turn the latter into a special juridical thing ('sacred' certain authors would venture to say – Labbée, 1990; 2002, p. 5), which no longer makes it possible to justify a difference in nature (in the legal sense) with animal bodies. On the other hand, there is a tendency to accept interventions on the human body (see, in particular, Thouvenin, 2007, p. 151) more willingly than in the past. Admittedly, we have already stated that the human body merges with the physical person during their lifetime and therefore benefits from a status that is totally different from that of animal bodies. However, the theories claiming that French law looks upon the body and the person separately have multiplied, and are linked directly to technical and scientific developments (Baud, 1993; 2007, p. 771). Since the corpse, the products of the body, the human embryo and the foetus do not fulfil the attribution criteria of legal personality, they come under the residual category of things. They belong to specific regimes and are not freely available,[6] but this characteristic on its own does not permit affirming human specificity. Animals too are objects of law governed by special rules strictly limiting the prerogatives of the owner, in the name of their quality as 'sensitive living beings'.

The fate reserved to 'spare' human embryos, initially created for the purpose of providing medically assisted procreation, but no longer needed to fulfil the parental project, is probably the example that most clearly demonstrates this development. The Bioethics Law of 1994 introduced the following statement in Article 16 of the *Civil Code*: 'The law [...] guarantees respect for the human being from the outset of his life'. Despite this, the constitutional judge refused to consider that this protection also applied to embryos created *in vitro* (Constitutional Council, 1994, jurisprudence p. 237, note B. Mathieu). The Bioethics Law subsequently made it possible, but only exceptionally, to carry out scientific experiments on these human embryos. In 2004 (*Loi n° 2004-800 du 6 août 2004*), the parliament created exceptions while maintaining the principle of interdiction. During the 2011 parliamentary debate focused on preparing a new version of the Bioethics Law (even though public debate had been going on since 2009), an amendment was proposed that would reverse the principle of interdiction. At the time, the amendment was rejected, but a new Law adopted in 2013 (*Law n° 2013-715 du 6 août 2013*) finally introduced a new legal framework. According to article L. 2151-5 of the French *Public Health Code*, scientific research on human embryos or embryonic stem cells is permitted under certain specific conditions and controlled authorisations. This modification does not allow for the free disposal of human embryos, even those considered 'spare'. Nor do the legal texts on the protection of animals permit the conclusion that experimenters have complete freedom. Animals, even before their birth, can only be used for experimental purposes under certain conditions. The legal texts concerning the protection of animals used for scientific purposes now also apply to certain invertebrates (cephalopods), as well as to autonomous larval forms and foetal forms of mammals, as of the last third of their development (Directive 2010/63/UE,[7] Article 1er, §3). Even the preceding phases can be included if the experiment implies allowing the animal to eventually continue its development and if it risks suffering.

Certain living human bodies – *in vitro* human embryos, for instance – are, consequently, legal objects for which technical-scientific interventions are now accepted. In parallel, interventions on the body of the human person have also become legal more frequently. In 2004 (*Loi n° 2004-800*), article 16-3 §1 of the *Civil Code* was also amended in order to authorise interventions on the human body 'in the event of medical necessity', and not only in the case of 'therapeutic' necessity. Obviously, this lexical change was introduced to broaden the range of legal exceptions. Consequently, interventions as diverse as sterilisation[8] and plastic surgery[9] – the latter of which opens the doors to 'body art' – are also covered. However, whether the aim is to prevent persons considered to be incapable of assuming parental responsibility from bearing children, or to enable others to try to be more attractive, these practices refer to enhancement, for individuals or for the human group (which does not imply the same consideration of the individual will). From a strictly legal point of view, it is therefore not iconoclastic to suppose that the principle of human specificity has been undermined.

Another question directly connected with 'enhancement' may once again reveal some weaknesses in the legal distinction between humans and animals. This is the issue of doping. Controlling the use of drugs is a problem commonly found in both human and animal races and sports competitions. The products used to strengthen resistance or physical capacities are, in effect, very similar, and the suppliers are frequently the same people (see, in particular, Berteau, 2000, p. 6; Defrance, 2004, p. 351).[10] The position of French law, in this instance, is identical for both practices: resorting to the use of drugs is forbidden and sanctioned. Specific legal measures have been devised to counteract those projects and techniques that aim to 'enhance' the body's performance.

Beyond the actual ban, does the common nature of the problems justify the same legal handling? The legislator has wavered between two options: should these practices be treated as one or as separate problems? Should synergies be created, improving the efficiency of the measures destined to fight against the circulation and use of doping products, or should specific measures focus on the health and individual care of human sports practitioners? In 1989, for example, a legal text (*Loi n° 89-432 du 28 juin 1989*[11]) globally treated the fight against doping in human as well as in animal competitions. However, this endeavour was received unfavourably in sports circles (see Lassalle, 1999). It came to a rapid end in practice, and then legally, through the adoption of the law of 23 March 1999 (*Loi n° 99-223*).[12] The new law only concerned the fight against doping and protection of human sports practitioners, in this way leading us to suppose that these objectives would be better served by specific measures and organisations with separate competences. But there was a further development in legislation in 2006 with the introduction of a law (*Loi n° 2006-405*)[13] taking the two phenomena into consideration and setting up a common institution to fight against doping, the French Agency for the Fight against Doping.[14] And yet the legislator did not create a single regime. The French *Sports Code*, for example, contains a Title III devoted to the 'health of sports practitioners and the fight against drug doping',[15] and a Title IV related to the 'fight against animal doping' (Articles L. 241-1 to L. 241-10). The measures concerning human sports practitioners take into account many developments in prevention and medical follow-up that are not included in the rules on doping in animals racing. Likewise, the committees that actually carry out the tasks of the Agency are specialised.

Doping, therefore, is one example of a complex legal situation, combining common health problems and comparable goals with a concern to maintain specific statements focused on the treatment dispensed to human beings. It can nonetheless be observed that the rules effectively introduced are not necessarily synonymous with the best protection for humans. Thus, it will be noted that there is a system of 'authorisation for therapeutic use' permitting a sick sports practitioner to participate in official competitions and events while receiving treatment. In principle, to do so, the sports practitioner in question must first obtain an authorisation from the Agency, backed by an official opinion of the Medical Committee. But in the case of a number of listed medical products, there is a lighter procedure that allows a doctor treating a sports practitioner to prescribe a product or a procedure, its authorisation being considered as obtained on receipt of the request addressed to the Agency. Some people see this as a genuine 'permission to take drugs'.[16] This point of view may appear to be excessive, but it should be noted that political determination to protect the health of sports practitioners reveals its shortcomings in this respect. Parliament, in fact, refused to introduce a rule forbidding a sick sports practitioner from competing, even though animals receiving veterinary treatment cannot participate in any competition, race or sports event. Their registration at the start of a race is subject to the delivery of a certificate declaring the absence of any trace of a medicinal product or one that is likely to have a doping effect. The rule for humans is thus the reverse of the rule for animals. The certificate in the latter case serves to check the health of the horse or the greyhound, but not to permit the sick animal to race under medication. Therefore, one can consider that the rule is more protective here for animals than for humans.

How can this paradoxical situation be explained? It would appear that economic issues exert an influence in both directions. On the one hand, they encourage horse owners to preserve the long-term health of an animal that will serve as a stud horse after being a champion (Bonnaire, 2000, p. 420). On the other hand, they incite organisers of human sports competitions to give priority to the show and the exploits of the players in the field, at the risk of health problems developing at a later stage (Desmoulin, 2006b, p. 852). It is tempting to link these facts to a linguistic drift that has been observed concerning contracts concluded in the area of sports, in which the idea that one can 'buy' players does not seem to shock anyone (the legal designation being 'transfer contract').[17] In a comprehensive (and somewhat provocative) manner, it can be observed that the principle of a different treatment for the bodies of humans and of animals has been mishandled for at least one of two reasons: either because a common treatment may perhaps be considered to be desirable, or because a different treatment leads concretely to less protection for human bodies than for animal bodies.

It is clear that the legal principle has been challenged by factual realities, in this case the physiological proximity of human and animal bodies and the similarity of the techniques used on both. Despite its vocation to introduce a coherent order in the organisation of life in society, the law is subject to the imperative (more or less justified and controlled) of keeping abreast of social developments. As a result, the legislator tries to adapt existing law by incorporating new factors (knowledge, practices, vocabulary), the implications of which are not always perceived. Most of the changes in the legal status of the human body, particularly human embryos created *in vitro*, or in the rules on doping follow this pattern. A topical example is the appearance of a new legal category: that of the 'human species'. This legal innovation was devised to prevent possible abuses of biotechnological applications. However, although the category of 'human species' is meant to raise an insuperable moral barrier between animals and humans, it is more likely to mark an additional stage towards a comprehensive legal approach to living entities (both human and non-human).

3 Confronting enhancement projects: does the law really protect mankind by protecting the 'human species'?

In 1999, the philosopher Peter Sloterdijk stirred up a controversy by publishing his lecture entitled 'Regeln für den Menschenpark' (translated as 'Règles pour le parc humain' in French and 'Rules for the human zoo' in English).[18] Deliberately resorting to concepts evoking 'domestication' or 'taming', in other words a vocabulary borrowed from the lexicon of breeders rather than that of educators, he proposed a reflection on the history and decline of literary humanism, prolonged by a questioning of the new possibilities of biotechnological interventions and their consequences for the future of the human species. He wrote at the time that:

> for the next period of time species politics will be decisive. That is, when it will be learned whether humanity (or at least its culturally decisive faction) will be able to achieve effective means of self-taming. [...] But, whether this process will also eventuate in a genetic reform of the characteristics of the species; whether the present anthropotechnology portends an explicit future determination of traits; whether human beings as a species can transform birth fatalities into optimal births and prenatal selection – these are questions with which the evolutionary horizon, as always vague and risky, begins to glimmer. (Sloterdijk, 1999c, p. 24)

Twelve years later, the parallel between animal selection and the prenatal screening of human embryos and foetuses (followed by selection in the case of *in vitro* embryos, or termination of pregnancies in the case of *in vivo* foetuses) is even more evocative. The announcement of the entry into post-humanism sounds strange in an era of 'transhumanist' declarations. According to authors of this movement, '[T]he human is but a complex component of matter [...] the fruit of a lengthy biological evolution [of which] it probably does not constitute the end' (French Transhumanist Association, 2009).[19] The human is thus repositioned in a *continuum* of living beings, contrary to the founding conceptual borders of French law. And yet French law displays a more contrasting image than ever before.

In fact, French legislators created a new legal category when drafting the Bioethics Laws of 1994 and 2004: the 'human species'. Thus, Article 16-4, §1 to 3, of the *Civil Code* states:

> No one may infringe upon the integrity of the human species ['espèce humaine']. Any eugenic practice aimed at organising the selection of persons is forbidden. Any procedure whose purpose is to cause the birth of a child genetically identical to another person, alive or dead, is forbidden.[20]

Similarly, the *Penal Code* sanctions eugenics and reproductive cloning (Articles 214-1 to 215-4 of the *Penal Code*) as crimes against the 'human species'.

By drafting specific legislation to protect the 'human species', legislators seem to have intended to promote humanist values. And yet in doing so, they have acknowledged the legal existence of a concept of the life sciences involving recognition of a biological *continuum*. According to Florence Bellivier (2007, p. 352), there is an unquestionable 'ambiguity in this: to protect the human species from technology, a biologising view of humanity is mobilised'. Henceforth, French law acknowledges the existence of offences committed against the protection of the human species just as it once acknowledged offences committed against the protection of endangered animal species. Admittedly, the applicable rules are not the same. Be that as it may, the parallel is no longer inappropriate. However, the same intellectual constructions are evoked to justify the special rules that have thus been elaborated: the concepts of '*patrimoine commun*' (common heritage) or '*choses communes*' (French adaptation of the Latin concept '*res communis*') are mobilised on the one hand (Labbée, 1999, Chronique p. 437; Chardeau, 2006), and '*sujet de droits*' (subject of law) on the other (Peis-Hitier, 2005, Chronique, p. 865).

Another piece of our puzzle has been added by the European Directive n° 2010/63/UE of 22 September 2010 on the protection of animals used for scientific purposes. Article 2, point 3, states that it applies to 'living non-human animals'. In view of this new line of thinking, it was indeed pertinent to state clearly whether reference is being made to human animals or not. The French administration in charge of adapting French regulation to the new directive decided to skip this definition and to translate the expression 'non-human primates' into 'great apes' ('*singes appartenant aux genres Gorilla, Pam et Pongo*') (*Decree n° 2013–118 of 1st February 2013*). Nevertheless, this step backward is not enough to thwart the tendency towards a merger and a clarification of the situation.

As for the protection of the 'human species', it should be noted that only certain human entities are protected. The offence of eugenics only applies to the selection of *persons*, which excludes human entities without a legal personality, and thus makes it legal to carry out prenatal screening and selection of the children to be born. This is why some scholars ask questions about the humanity of the embryo *in vitro* (Fédida et al., 1996; Herzog-Evans, 2000, p. 65), while others, in a more or less provocative manner, talk of 'improvement of the quality of the "product"', when referring to children born as a result of new reproductive techniques (Fagot-Largeault, 2011). This tendency does not seem to have been reversed since 2004: a report published by the Agency of Biomedicine[21] seems in favour of broadening the range of diseases for which access to preimplantation genetic diagnosis before embryo transfer into the uterus of a future mother is allowed (Agence de la Biomédecine, 2008).

Article 16-4 of the *Civil Code* and the new incriminations concerning the protection of 'the human species' were intended, according to the parliamentary debates, to combat projects such as those of the 'extropians' (see Le Breton, 1999, pp. 212 and following) or transhumanists. Transhumanism is an international, cultural and intellectual movement advocating the use of science and technology to 'enhance' the physical and mental characteristics of human beings, with the aim of eradicating suffering, handicaps and the pernicious effects of ageing (Fukuyama, 2004; Besnier, 2009). Its supporters rely on Darwin's theory of evolution to demonstrate that humanity, as it exists, is not the final point of biological evolution, but just one phase (Bostrom, 2005). They hope that biotechnologies, nanotechnologies and technologies of information and cognition (the Nano-Bio-Info-Cogno 'convergence') will provide the means for guiding human evolution towards an objective of 'enhancement' that they feel is eminently desirable. All procreative techniques (including cloning) that make it possible to control the health and performance of one's offspring, and all interventions on the body (including transgenesis) are perceived as tools at the service of this goal. The 'enhancement' thus sought is, in actual fact, a surpassing of humans as we know them today. Transhumanists call for a breaking away from the straightjacket of current thinking in order to conceive a 'new' humanity. This would make it possible to imagine 'enhancing' humans by creating bodies that borrow features partly from the 'old humanity', and partly from other animals. Drawing inspiration from the arguments put forward by the utilitarian current, Article 7 of the Declaration of

Transhumanist Rights 'advocates the welfare of all living creatures with feelings, whether they come from a human, artificial, post-human or animal brain' (Transhumanist Declaration, 1998). Analysing this declaration, Marie-Angèle Hermitte (2011, p. 157) notes that 'freedoms – of research, experimentation and intervention on oneself and one's offspring – are the necessary ingredients for the transhumanist project, as a negation of all ideas of human nature'. With this in mind, the idea of human specificity, based in particular on the legal status of human persons and human bodies, no longer has any meaning. All that remains are the references to the individual will, to the capacity to suffer and the imperative to combat the physical and moral suffering imposed on the human body by its present limitations.

Despite the flaws described previously, the French laws aimed at protecting the 'human species' forbid eugenics and cloning (reproductive and therapeutic). By doing so, they reject some of the principal means demanded by transhumanists to implement their projects. However, the durability and effectiveness of these measures in the long run could be challenged. In addition to the gradual modifications already mentioned, and to the possible changes concerning prenatal testing of diseases or therapeutic cloning, a major conceptual weakness must be pointed out. The biologising view that lies at the core of the new legal category of 'human species' could undermine the legal structure. It is in fact laborious to use such a conceptual term without including an implicit reference to evolutionism. If that is the case, the legal affirmation of the 'integrity' of the human species becomes so difficult to interpret that it can raise serious doubts as to the effectiveness of the law. How can permanent change and a determination to maintain the present status be reconciled? Furthermore, transhumanist theories are actually based on a certain notion of evolution. In this context, Florence Bellivier (2007, p. 354) writes that 'it is conceivable that the contested applications of techno-science are a mere drop in the ocean of the evolution of the species – which renders the concept of integrity rather vague'.

Do the experiments carried out or planned on embryonic chimeras, cybrids and other experimental mixtures of human and animal cells constitute the first step towards a gradual lack of differentiation between human and animal bodies? This theory is plausible during the laboratory stage. Such research has already been carried out on the other side of the Channel and of the Atlantic (Shreeve, 2005, p. 5; Green et al., 2005, p. 385; Coghlan, 2007). The reactions of legal systems to these activities vary from one country to another (Taupitz and Weschka, 2009). In France, where research on animal-animal chimeras is highly developed (Le Douarin, 2000), scientists and the Agency of Biomedicine have remained discreet about the idea of discovering how to improve the human machinery by studying embryogenesis based on human-animal cell mixtures. The 2011 version of the Bioethics Law states that the creation of chimerical embryos is forbidden (Article L. 2151-2, §2, of the French *Public Health Code*).[22] These research studies may provoke fears about the potential creation of 'monsters', as they did in the era of the surgeon and anatomist Ambroise Paré (1510–90) and the philosopher and physician Fortunio Liceti (1577–1657) (Roger, 1987, pp. 31–90), but from the transhumanist angle, on the contrary, they raise hopes about the 'improvements' that could be made to the human body. This vision, however, does not stop at the borders of the living. The next stage – which apparently is already under way with research on neuronal implants, human-computer interfaces and artificial intelligence – is hybridisation with machines. Raising the performance or attractiveness of bodies, increasing their cognitive capacities through connections with machines and even changing bodies, contribute, through the selection and manipulation of gametes and embryos, to what some philosophers now refer to as *procreative beneficence* (Savulescu, 2001; see also Buchanan et al., 2000, 'Scenario 5 the genetic enhancement certificate', p. 156 and following). But here we are overstepping the limits of our study.

Notes

1. All translations of French sources are mine, except for French law, for which I have provided an official translation whenever it is available (see: http://www.legifrance.gouv.fr/Traductions/en-English).
2. See chapters by F. Burgat and A. Ferrari.
3. See, in particular, Jouannet (2007, p. 767): 'Since man is a mammal among others, at least from the biological point of view, it is not surprising that the techniques perfected for other species can be applied to humans. This is what has led to the medicalisation of procreation'.
4. Stemming from *Ordonnance n° 2006-1548 du 7 décembre 2006*.
5. Decree n° 2007-818 of 11 May 2007 related to health approval of animal reproduction activities and health regulations concerning these activities and modifying the *Rural Code*.

6   Article 16-1, §2 and 3 of the *Civil Code*: 'The human body is inviolable. The human body, its elements and its products may not be the object of an inheritance right'.
7   Directive 2010/63/UE of 22 September 2010 on the protection of animals used for scientific purposes.
8   Since 2001, Articles L. 2123-2 and following of the *Public Health Code* make provisions for the 'sterilisation for contraceptive purposes' of persons suffering from a deterioration of mental capacities causing a handicap.
9   Articles L. 6322-1 and following of the *Public Health Code* cover aesthetic surgery.
10 For an example of jurisprudence, see *Cour de Cassation*, Criminal Chamber, 14 March 2006, Appeal n° 05-87791, which deals with the question of supplying drug substances to a team of boxers and racing horse trainers.
11 *Loi n° 89-432 du 28 juin 1989* on 'the prevention and repression of the use of drug products on the occasion of sports competitions and events', amended in 1999, became the law on 'the repression of doping of animals participating in sports competitions'.
12 *Loi n° 99-223 du 23 mars 1999* on 'protecting the health of sports practitioners and the fight against doping'.
13 *Loi n° 2006-405 du 5 avril 2006*, 'dealing with the fight against doping and the protection of the health of sports practitioners'. Two later texts adopted in 2010 (*Ordonnance n° 2010-379 du 14 avril 2010*) and 2012 (*Loi n° 2012-158 du 1er février 2012*) modified the legal requirements without changing the legal solutions described here.
14 Article L. 232-5 of the *Sports Code*: 'The French Agency for the Fight against Doping, an independent public authority with a moral personality, defines and implements actions to fight against doping. For this purpose, it cooperates with the World Anti-Doping Agency and with international sports federations'. Article L. 241-1 of the same Code: 'The French Agency for the Fight against Doping defines and implements the actions set forth in Article L. 232-5 to fight against animal doping'.
15 The measures devoted to the fight against doping are contained in Articles L. 232-1 to L. 232-31 of the *Sports Code*.
16 See the minutes of the parliamentary debates on the bill on doping, session of 19 October 2005, Senate, and session of 23 March 2006, AN. Adde Lapouble (2006).
17 For examples of jurisprudence on 'transfer contracts' between football clubs and the difficulties of implementation: Cour de Cassation, Commercial Chamber, 2010; Cour d'Appel de Douai, 2010.
18 Originally published in German in 1999, in the newspaper *Die Zeit* (Sloterdijk, 1999a), the text was translated into French the same year for a news magazine (Sloterdijk, 1999b) and republished 10 years later in English for a journal (Sloterdijk, 2009a) and in French for a book (Sloterdijk, 2009b).

19 'Cahier d'acteur' submitted to the National Commission of Public Debate by the French Transhumanist Association during the debate on nanotechnologies, 2009: http://www.debatpublic-nano.org/documents/liste-cahier-acteurs.html.
20 The translation proposed by the Ministry of Justice's website *Légifrance* refers to 'the integrity of *mankind*', but this translation does not respect the original version of the legal text and the intention of the legislator.
21 The Agency of Biomedicine (Agence de la Biomédecine) is the authority concerned with embryology, procreation and human genetics in France.
22 *Loi n° 2011-814 du 7 juillet 2011 relative à la bioéthique.*

References

AFSSA (2005) *Bénéfices et risques liés aux applications du clonage aux animaux d'élevage* (Paris: AFSSA).
Agence de la Biomédecine (2008) *Bilan d'application de la loi de bioéthique du 6 août 2004. Rapport à la Ministre de la santé, de la jeunesse, des sports et de la vie associative* (Paris: Agence de la Biomédecine, October 2008).
Antoine S. (2005) *Rapport sur le régime juridique de l'animal* (Paris: Ministry of Justice, 10 May).
Article 16-1, §2 and 3, *Civil Code*.
Articles 214-1 to 215-4, *Penal Code*.
Articles 214-2, 511-1-2, 511-18-1, 517 and 518, *Penal Code*.
Articles L. 2123-2 and following, L. 2151-2, L. 6322-1 and following, *Public Health Code*.
Articles L. 653-1 and following, *Rural Code*.
Articles L. 232-1 to L. 232-31; L. 241-1 to L. 241-10, *Sports Code*.
Bateman S. (2004) 'La nature fait-elle (encore) bien les choses?', in Pharo P. (ed.), *L'homme et le vivant* (Paris: PUF 'Science, histoire et société'), pp. 391–404.
Baud J.-P. (1993) *L'affaire de la main volée. Une histoire juridique du corps* (Paris: Le Seuil), 243 p.
Baud J.-P. (2007) v° Propriété, in Marzano M. (ed.), *Dictionnaire du corps* (Paris: PUF 'Quadrige'), pp. 770–6.
Bellivier F. (2007), v° Espèce humaine, in Marzano M. (ed.), *Dictionnaire du corps* (Paris: PUF 'Quadrige'), pp. 351–4.
Berteau P. (2000) 'Dopage, droit et médecine du sport', *Médecine et Droit*, 44 (Paris: Elsevier), pp. 6–15.

Besnier J.-M. (2009) *Demain les posthumains. Le futur a-t-il besoin de nous?* (Paris: Fayard 'Haute tension'), 216 p.
Bonnaire V.Y. (2000) 'Le dopage de l'animal de compétition', in P. Laure (ed.), *Dopage et société* (Paris: Ellipses), pp. 420–32.
Bostrom N. (2005) 'A history of transhumanist thought', *Journal of Evolution & Technology*, 14, April 2005, http://jetpress.org/volume14/freitas.html.
Buchanan A., Brock D.W., Daniels N. and Wikler D. (2001) *From Chance to Choice. Genetics and Justice* (Cambridge: CUP), 414 p.
Carbonnier J. (2004) *Droit civil, 1, Introduction. Les personnes. La famille, l'enfant, le couple* (Paris: PUF 'Quadrige'), pp. 373–418.
Cassin B. (ed.) (2004), v° Comparaison, *Vocabulaire européen des philosophies* (Paris: Le Seuil-Robert), pp. 243–8.
Chardeau M.-A. (2006) *Les choses communes* (Paris: LGDJ 'Bibliothèque de droit privé'), t. 464, 504 p.
Coghlan A. (2007) 'Frankenbunny, human, or cybrid?', *New Scientist*, 5 January, http://www.newscientist.com/blog/shortsharpscience/2007/01/frankenbunny-human-or-cybrid_05.html.
Constitutional Council (1994) '27 July 1994', in *Recueil Dalloz* (Paris: Dalloz) 1995, Jurisprudence p. 237, note B. Mathieu.
Cour d'Appel de Douai (2010) 'Appeal n° 09/05120, SASP Sté Stade Malherbe de Caen Calvados Basse-Normandie c/ SASP LOSC Lille Métropole', 16 September.
Cour de Cassation, Commercial Chamber (2010) 'Appeal n° 09-65805, SASP ASSE Loire c/ Sté Club de football Zénit', 1 June.
Cour de Cassation, Criminal Chamber (2006) 'Appeal n° 05-87791', 14 March.
Defrance J. (2004) v° Dopage, in D. Lecourt (ed.) *Dictionnaire de la pensée médicale* (Paris: PUF), pp. 350–2.
Desmoulin S. (2006a) *L'animal, entre science et droit* (Aix-en-Provence: Presses Universitaires d'Aix-Marseille), §192–204.
Desmoulin S. (2006b) 'Lutte contre le dopage et encadrement médicalisé des activités sportives. Remarques à propos de la loi n° 2006-405 of 5 April 2006', *Revue de droit sanitaire et social*, 5, October, 852–64.
Directive 2010/63/UE of 22 September 2010 on the protection of animals used for scientific purposes.
EFSA (2008) 'Scientific Opinion of the Scientific Committee, Food Safety, Animal Health and Welfare and Environmental Impact of

Animals derived from Cloning by Somatic Cell Nucleus Transfer (SCNT) and their Offspring and Products Obtained from those Animals', *EFSA Journal*, 767, 1–49.

EFSA (2009) 'Statement – Further Advice on the Implications of Animal Cloning (SCNT)', *EFSA Journal*, RN 319, 1–15.

EFSA (2012) 'Statement – Update of the State of Animal Health and Welfare and Environmental Impacts of Animals derived from SCNT Cloning and their Offspring, and Food Safety Products Obtained from those Animals', *EFSA Journal*, 10(7), 2794.

Fagot-Largeault A. (2011) 'Les nouveaux modes de procréation', in *Hominisation, humanisation: le rôle du droit* (Paris: Workshop at the Collège de France), 29 April, http://www.collegedefrance.fr/default/EN/all/int_dro/Seminaire_du_29_avril_2011_Hom.htm.

Fédida P., Lecourt D., Mattei J.-F. et al. (1996) *L'embryon humain est-il humain?* (Paris: PUF 'Forum Diderot'), 93 p.

French Transhumanist Association (2009) 'Cahier d'acteur', http://www.debatpublic-nano.org/documents/liste-cahier-acteurs.html.

Fukuyama F. (2004) 'Transhumanism', *Foreign Policy*, 1 September.

Green M. et al. (2005) 'Ethics: moral issues of human-non human primate neural grafting', *Science*, 309.

Hermitte M.-A. (2011) 'De la question de la race à celle de l'espèce – Analyse juridique du transhumanisme', in Canselier G. and Desmoulin-Canselier S. (eds), *Les catégories ethno-raciales à l'ère des biotechnologies. Droit, sciences et médecine face à la diversité humaine* (Paris: Société de Législation comparée 'UMR de droit comparé'), pp. 155–70.

Herzog-Evans M. (2000) 'Homme, homme juridique et humanité de l'embryon', *Revue trimestrielle de droit civil*, 65–78.

Jouannet P. (2007) v° Procréation médicalisée, in Marzano M. (ed.), *Dictionnaire du corps* (Paris: PUF 'Quadrige'), pp. 766–70.

Labbée X. (1990) *La condition juridique du corps humain avant la naissance et après la mort* (Lille: Presses universitaires de Lille), rééd. Presses universitaires du Septentrion, 2012, p. 450

Labbée X. (1999) 'Esquisse d'une définition civiliste de l'espèce humaine', in *Recueil Dalloz* (Paris: Dalloz), pp. 437–42.

Labbée X. (2002) 'La personne, l'âme et le corps', *Les Petites Affiches*, 243, 5 December, 5–8.

Lapouble J.-C. (2006) 'La lutte contre le dopage et la protection de la santé des sportifs', *JCP-La Semaine Juridique* (Paris: Juris-Classeur), édition générale 2006, I, 136, n° 17, 893–8.

Lassalle J.-Y. (1999) 'Le dopage des sportifs: une nouvelle loi', *JCP-La Semaine juridique* (Paris: Juris-Classeur), édition générale, I, 133, 845–52.
Le Breton D. (1999) *L'adieu au corps* (Paris: Métailié), 238 p.
Le Douarin N. (2000) *Des chimères, des clones et des gènes* (Paris: Odile Jacob 'Sciences'), 496 p.
Le Monde (1998) 'Clonage: le gouvernement néerlandais interdit le clonage de bovins', *Le Monde*, 2 March.
Loi n° 89–432 of 28 June 1989.
Loi n° 99–223 of 23 March 1999.
Loi n° 2004–800 of 6 August 2004.
Loi n° 2006–405 of 5 April 2006.
Loi n° 2011–814 of 7 July 2011.
Loi n° 2013–715 of 6 August 2013
Marguénaud J.-P. (1998) 'La personnalité juridique des animaux', in *Recueil Dalloz* (Paris: Dalloz), Chronique p. 205.
Peis-Hitier M.-P. (2005) 'Recherche d'une qualification juridique de l'espèce humaine', in *Recueil Dalloz* (Paris: Dalloz), Chronique, pp. 865–9.
Roger J. (1987) *Les sciences de la vie dans la pensée française au XVIIIe siècle* (Paris: Albin Michel 'L'évolution de l'humanité'), 852 p.
Savulescu J. (2001) 'Procreative beneficence: why we should select the best children', *Bioethics*, 15(5/6), 413–26, http://www.blackwellpublishing.com/content/BPL_Images/Journal_Samples/BIOT0269-9702~15~5&6~251%5C251.pdf.
Shreeve J. (2005) 'L'avenir des chimères', *Futuribles*, 312, October.
Sloterdijk P. (1999a) 'Regeln für den Menschenpark', *Die Zeit*.
Sloterdijk P. (1999b) 'Règles pour le parc humain. Réponse à la Lettre sur l'humanisme', French translation of 1999a, *Le Monde des débats*, October, additional document.
Sloterdijk P. (2009a) 'Rules for the human zoo: a response to the Letter on humanism', English translation of 1999a, *Environment and Planning D: Society and Space*, 27, 12–28.
Sloterdijk P. (2009b) 'Règles pour le parc humain. Réponse à la Lettre sur l'humanisme', French translation of 1999a, in *'Règles pour le parc humain' suivi de 'La domestication de l'être'* (Paris: Mille et une nuits).
Taupitz J. and Weschka M. (2009) *Chimbrids – Chimeras and Hybrids in Comparative European and International Research: Scientific, Ethical, Philosophical and Legal Aspects* (Berlin: Springer-Verlag), 1058 p.

Thouvenin D. (2007) 'Le corps à corps', in David-Ménard M. (ed.), *Autour de Pierre Fédida. Regards, savoirs, pratiques* (Paris: PUF), 151–75.

Transhumanist Declaration (1998, modified version 2009), http://humanityplus.org/learn/transhumanist-declaration/.

Index

ableism, 24–5
abolitionism, 57–8, 64–5, 68
abolitionist project, 15, 23, 27n9
abolitionists, of animal research, *see* animal rightists
Ali, A., 63–4
American Breeders' Association (ABA), 21
animal breeding, 3, 4, 5, 6, 16, 18–19, 21, 27n5, 35, 37–9, 111
  lessons from, 6–7, 8–11, 52–7
  see also animal enhancement; zootechnics
Animal Breeding Plans, 21, 38
animal disenhancement, 6, 15–16, 23
animal enhancement, 2–4
  advocates (transhumanist) of, 17–18, 20, 21–2, *see also* transhumanism
  in agriculture, 14–15, 16, 19, 21, 27n6, 58–9, 83–7
  and augmentation, 81, 85, 93, 103
  for cognitive function improvement, 14, 26n3, 59–62
  as colonization of nature, 23–5
  compared to human enhancement, 4–5, 6, 7–8, 10, 16, 40–4, 112–16
  for competitions (sports), 22, 27n6, 40, 118, 119, 120, 126n11
  effects of, 17–20, 35, 53–4, 60–1, 63–4
  and ethics, 15, 18, 20–1, 25, 28n17, 43–4, 47n3, 54, 56, 58, 65, 68, 70–4
  humans as drivers of, 4–5, 8, 18–23, 27n11, 52–3, 81
  lessons from, 6–7, 8–11, 52–7
  purposes of, 2–3, 4–5, 7, 9, 17–20
  and sex hormones, 80–2, 89, 95, 98, 99, 101, 102
  in sports, 14, 17, 19, 22, 25, 27n6, 118
  techniques/practices, 5, 7–9, *see also* animal experimentation; animal research; enhancement, techniques/practices
  as technovisionary paternalism, 20–3
  see also animal improvement
animal experimentation, 4, 6–7, 11, 14, 41, 50, 57
animal models, 17, 50, 57, 73, 92, 93, 104n9
  effects of, 17, 19, 23, 26–7n4, 35, 53–4, 61, 63–4
  silence about, 17–20, 26–7n4
  see also transgenic animals

animal exploitation, 6, 15, 22, 23, 25–6
animal husbandry, 6, 22, 25, 52, 53, 80, 96, 99
animal improvement, 3–6, 8–9, 14–15, 35
  comparison with human improvement, 40–5
  objectives of, 36, 41, 43, 44–5
  origin of, 37–8
  problems raised by, 35, 38, 40–1
  techniques/tools for, 37–8, 41, 44–6
  and well-being, 39–40, 41, 43, 47n3
  see also animal enhancement; enhancement, techniques/ practices; zootechnics
animal models, 17, 50, 57, 73, 92, 93, 104n9
animal research
  abolitionism, 57–9, 64–5, 68
  abolitionists of, 50, 57–9, 62, 64
  policies that regulate, 70–1, 72–3, 75n3
  reformists of, 58–9, 68–71
  speciesism, 57, 65–8
  speciesists of, 50, 57, 65–8
  two-level utilitarians of, 68–71
  see also Animal Welfare Act (AWA)
animal rightists, 50, 57–9, 62, 64
animal rights, 7, 8–9, 11, 17–18, 19, 26, 74
  abolitionist understanding of, 57–8, 63–4
  and ethical review, 72
  and informed consent, 72
  reformist view of, 58–9, 68–71
  speciesist conclusions of, 65–8
  utilitarian understanding of, 50, 68–71, 73
  and the worse-off principle, 62–3, 68, 72
  see also Animal Welfare Act (AWA)
animal selection, 2–3, 36, 121
animal welfare, 6, 16, 19, 23, 64, 68, 102
  and ethical review, 72
  and informed consent, 72

laws, 50, 70–1, 73–4, 75n3
  see also Animal Welfare Act (AWA)
Animal Welfare Act (AWA), 70–71, 73, 75n3
  rules of, 70–1, 73–4
  three Rs, 71, 73
animal(s)
  as companions, 5, 9, 52
  diminishing of abilities, 6, 15–16, 19, 23, 60–1
  experimental, 7, 10, 16, 51, 52, 67, 70, 72, 74
  farm, 6, 19, 21, 25, 37, 71, 81, 99
  and food production, 3, 4, 5, 9, 88, 101
  genetic testing for, 19–20
  human utility of, 5, 6, 7, 9, 17–19, 25, 45–6, 51–2
  hyper-types, 40
  laboratory, 10, 16, 35, 90, 99
  mass killing of, 19, 28n16
  and moral rights, 66–7
  non-human, 5, 11, 26n2, 27n11, 28n22, 45, 65, 122
  as pets, 6, 27n7, 35, 40, 41
  protection of rights/interests, 8, 11, 17–18, 23, 72
  rights of, *see* animal rights
  as 'subjects-of-a-life', 7, 10–11, 57–8
  sufferings of, 17, 19, 23, 26–7n4, 35, 53–4, 61
  transgenic, *see* transgenic animals
artificial insemination, 5, 37, 41, 112, 114
artificial selection, 3, 19
  see also genetic selection
augmentation, 8, 81, 85, 93, 103

Bateman, S., 114
Baudement, E., 37, 38
Bell, S., 82–3
Bellivier, F., 122, 124
bioethics, 15, 111, 112, 116, 117, 122, 125
biotechnologies, 3, 15, 37, 41, 47n8, 54, 110, 111, 123
Bostrom, N., 2, 26n1, 54

breeding, 2, 6, 18–19, 20–1
 see also animal breeding; industrial breeding
Burroughs, W., 83

Carbonnier, J., 112
Carson, R., 85
cattle
 bred for meat, 40–1
 DES use in, 7, 80, 81, 83, 88–9, 91, 94–8
 improvement of, 15, 36, 37, 38, 52
 mastitis in, 14, 19, 28n16, 53
Chan, S., 6, 14, 15, 16, 17, 21
The Chemical Feast, 85
Cheng, K. M., 63–4
chimpanzees, 15, 22–3, 25
Citizen Cyborg, 15
civil law systems, 4, 5, 7–8, 10, 111–16, see also Roman law
cloning, 5, 14, 41, 46, 52, 123, 124
 animal, 114–15
 French laws on, 114–16, 122
 human, 115–16
 reproductive, 5, 115, 122, 124
 sport horses, 17, 22, 27n6
 therapeutic, 116, 124
 see also somatic cell nuclear transfer (SCNT)
CNAG, see National Commission for Genetic Improvement (*Commission Nationale pour l'Amélioration Génétique*, CNAG)
cognitive enhancement, 14, 26n3, 59–62
Cohen, C., 66, 67
competitions (sports), 22, 27n6, 40, 118, 119, 120, 126n11
cows, 14, 19, 28n16, 36, 53, 89

Dagognet, F., 35, 47n8
Delaney clause (1958), 84–5, 104n11
de-vocalisation, 18
Diethylstilbestrol (DES), 80–104
 American versus French approach to, 94–103

 for animal augmentation, 81, 85, 93, 103
 ban on, 80, 84, 91–2
 carcinogenic effects, 83, 85–7, 88–90, 98
 -daughters, 80, 82, 87, 95, 101
 as growth enhancer in cattle, 7, 80, 81, 83, 88–9, 91, 94–8
 medical uses, pregnant women, 7, 80–81, 82–3, 88, 94, 100
 regulations on use of, 82, 84, 85–7, 88, 97–8, 100–1
 as synthetic oestrogen, 7, 81, 82–3, 89–90
Dodds, C., 82
dogs, 15, 18, 27n7, 37, 40, 53
domestication, 37, 52, 121
doping, 2, 8, 22, 118–20, 126nn11–16
Dvorsky, G., 15

ear cropping, 18
Edwards, C., 89, 92, 93
embryos (human)
 cryopreservation of, 37, 41, 114
 genetic screening of (as compared with animal selection), 121
 research on, 117–8
enhancement
 of humans as compared to animals, 4–5, 6, 7–8, 10, 16, 40–4, 112–16
 techniques, see enhancement techniques/practices
 see also animal enhancement; animal experimentation; genetic enhancement; growth enhancement; human enhancement; performance enhancement
enhancement, regulation of
 administrative, 91, 93–4, 96, 100
 and civil law systems, 4, 5, 7–8, 10, 111–16
 congressional hearings (US), 82, 87–92, 93, 102
 DES, regulations on use of, 82, 84, 85–7, 88, 97–8, 100–1

enhancement – *continued*
  legal, 82–3, 84–5, 88
  regulatory capture, 80, 93
  see also laws; laws (France)
enhancement techniques/practices, 5, 7–9
  goals and values of, 5, 9, 10
  in humans versus animals, 112–16
  techniques/tools for, 37–8, 41, 43–5
  technological interventions, criteria for, 6, 16
  as 'therapy' in humans, 5, 8
  see also artificial insemination; biotechnologies; breeding; cloning; Diethylstilbestrol (DES); doping; genetic engineering; hybridisation; reproductive technology; selective breeding; transgenesis
Enviropigs, 14, 19, 25, 27n5, 53
ethics, 15, 18, 20–1, 25, 28n17, 43–4, 47n3, 54, 56, 58, 65, 68, 70–4
  see also bioethics
eugenics, 2, 5, 10, 20–1, 42, 122, 123, 124
experimentation, *see* animal experimentation; human experimentation

Faure, J.-M., 39
Ferrando, R., 98–9
Food and Drug Administration (FDA), 80–5, 87–94, 96, 100, 101
  Delaney clause (1958), 84–5, 104n11
  Food and Drug Act (1962), 85, 88
  Food, Drug, and Cosmetic Act (1938), 82–3
  see also Diethylstilbestrol (DES)
Fountain, L., 88–9, 93

gene therapy, 15, 17
genetic engineering, 14, 16, 17, 19, 55, 63
genetic enhancement, 5, 9–10, 15, 50–1, 55, 57, 59, 62, 63, 64, 71, 74, 111, 125
genetic improvement, 38, 41, 46, 113
genetic modification, 2, 9, 21, 39
genetic selection, 20, 28n17, 44–5, 112
genetic testing, 18, 19–20

Hare, R. M., 69, 72
Herbst, A., 86–7, 88, 89, 92, 94
horses, 3, 8, 17, 22, 37, 53, 120
Hughes, J., 15
human enhancement
  compared to animal enhancement, 4–5, 6, 7–8, 10, 16, 40–4, 112–16
  concept of, 14
  DES and, *see* Diethylstilbestrol (DES)
  lessons for human enhancement, 6–7, 8–11, 52–7
  and protection of rights/interests, 8–9, 11, 25
  purposes of, 7, 9–10
  as therapy/treatment, 5, 8
  see also enhancement, techniques/practices
human experimentation, 11, 42
human improvement, 35
  comparison with animal improvement, 40–5
human rights, 11, 65–9, 73
human(s)
  /animal divide, 8–9
  as continuum of living beings, 10, 121–2
  DES use and cancer in, 80, 83, 85–7, 88–90, 98
  as drivers of animal enhancement, 4–5, 8, 18–23, 27n11, 52–3, 81
  see also human enhancement; human experimentation; human improvement; human rights
hunting, 19, 43, 52–3, 58
hybridisation, 3, 9, 125

improvement, *see* animal improvement; genetic improvement; human improvement
industrial agriculture, 80, 83–4, 85, 86, 96, 102
industrial breeding, 42, 43, 47n3, 47n8
INRA, *see* National Institute for Agricultural Research (*Institut National de la Recherche Agronomique*, INRA)

integrity
 of animals, 43, 54, 64
 of human species, 122, 124
intentional genetic enhancement of animals (IGEA), 57, 63, 65, 68
interests
 animal, 4, 6, 8–9, 11, 14, 16, 17, 18, 38, 46, 60, 72
 economic, 28n12, 80, 86, 102
 human, 11, 16, 21, 46, 69, 74
 owner, 4, 18

laws, see animal welfare; laws (France); Roman law
laws (France)
 on breeding, 36, 38–9
 Bioethics Laws, 111, 112, 116, 117, 122, 125
 civil law system, 4–5, 8, 112–13
 on cloning, 114–16, 122
 on DES use, 97–8
 on doping, 8, 118–20
 on enhancement techniques in humans vs animals, 7–8, 112–16
 on human embryos, 117–18, 120, 121, 123, 125
 on persons and things, 4–5, 10, 111–12
 on protection of human species, 121–5
 on reproductive technologies, 113, 122
lessons for human enhancement, 6–7, 8–11, 52–7
Lipsett, M., 89
Longgood, W., 85
Lush J. L., 21, 38

Mäntyranta, E., 50, 51
Marcus, A., 83–4, 86
McMahan, J., 66, 67
Meyers-Wallen, V. N., 20, 28n17
Mills, A., 39
Minvielle, F., 36
moral status, 11, 28n22, 65, 67

Nader, R., 85, 91, 97

National Academy of Sciences (NAS), 88
National Cancer Institute (NCI), 91, 92
National Commission for Genetic Improvement (*Commission Nationale pour l'Amélioration Génétique*, CNAG), 38–9
National Institute for Agricultural Research (*Institut National de la Recherche Agronomique*, INRA), 38, 39–40
nature, 20, 22, 23–6, 43, 90
non-human animals/beings, 5, 10–11, 15, 21, 22, 26n2, 27n11, 28n22, 45, 65, 72, 121, 122

oestrogens, 7, 81–4, 89–91, 95–8, 99
see also Diethylstilbestrol (DES)

Pearce, D., 15, 17, 21, 23, 26
performance, see performance enhancement/improvement
performance enhancement/ improvement, 3, 8, 14, 15, 22, 24, 36, 41, 44, 46, 51, 59–62, 99, 110, 115, 118, 125
see also doping
pigs, 14, 19, 27n5, 35, 42, 52
see also Enviropigs
The Poisons in Your Food, 85
protection of rights/interests, 8–9, 11, 18, 23, 72, 73–4
see also animal rights; human rights

Raven, P. G., 22–3
reformists, of animal research, 58–9, 68–71
Regan, T., 57–8, 62, 65, 75n2
regulations, see enhancement, regulation of
Rekha, J., 59–62
reproductive revolution, 15, 27n9
reproductive technology, 10, 17, 37, 41, 113–15, 123, 124
Rise of the Planet of the Apes (movie), 14

Roman law, 4–5
 distinction between persons and things, 4–5, 8, 10–11, 111–12
 see also civil law systems; laws (France)

Savulescu, J., 15, 17, 21, 26n1, 28n22, 125
selection, *see* animal selection; artificial selection; genetic selection
selective breeding, 6, 9–10, 21
sex hormones, 7, 80–2, 89, 95, 98, 99, 101, 102
 see also Diethylstilbestrol (DES)
Silent Spring, 85
Singer, P., 23, 58
Sloterdijk, P., 121, 126n18
Smith, J., 83
somatic cell nuclear transfer (SCNT), 116, *see also* cloning
Spallanzani, L., 37
speciesism, 57, 65–8
 speciesists, of animal research, 50, 57, 65–8
Streiffer, R., 53, 61
'subjects-of-a-life', 7, 10–11, 57–8

tail docking, 18, 28n13, 28n14
Tanguy-Prigent, F., 38
technological interventions, 3, 14, 24, 121
 criteria for qualification as enhancement, 6, 16
 for non-therapeutic reasons, 18
transgenesis, 3, 5, 14, 41, 46, 123
transgenic animals, 14–15, 17, 46
 chickens, 63–5
 cows, 14, 19, 28n16, 36, 53, 89
 dogs, 14, 53
 goats, 14
 mice, 14, 28n18, 59–62
 monkeys, 62–3
 rats, 52, 53–4
 pigs, *see* Enviropigs
transhumanism, 2, 22, 23–4, 123
transhumanists, 6, 10, 15, 20, 22, 28n18, 54, 121, 123–4, 125, 127n19
Turner, J. S., 85
two-level utilitarianism, 50, 68–71, 73

uplift biotechnologies, 15
U.S. Department of Agriculture (USDA), 91, 92

Varner G. E., 64, 68, 69
veterinary medicine
 expertise, 90, 97–8
 interventions, 18, 28n12, 39, 119

worse-off principle, 62–3, 68, 72

zootechnics, 36, 37, 38, 39, 42, 43, 45